G

DEC 29 2010

Robert Benewick is Professor Emeritus in the Department of Politics at the University of Sussex. His recent publications include *Asian Politics in Development* with Marc Blecher and Sarah Cook (eds.) and articles on community construction in China.

Stephanie Hemelryk Donald is Professor of Chinese Media Studies at the University of Sydney. An expert on the politics of media and culture and the intersection of social history and visual media, she is author of *Public Secrets, Public Spaces* (2000), *Little Friends* (2005), *Tourism and the Branded City: Film and Identity on the Pacific Rim* (2007), *Media in China* (co-edited with Michael Keane, 2001) and *Picturing Power* (co-edited with Harriet Evans, 1999). She has served as the president of the Chinese Studies Association of Australia and as Chair of the Humanities and Creative Arts panel of the Australia Research Council.

D0128582

"Unique and uniquely beautiful.... A single map here tells us more about the world today than a dozen abstracts or scholarly tomes." *Los Angeles Times*

"A striking new approach to cartography.... No-one wishing to keep a grip on the reality of the world should be without these books." *International Herald Tribune*

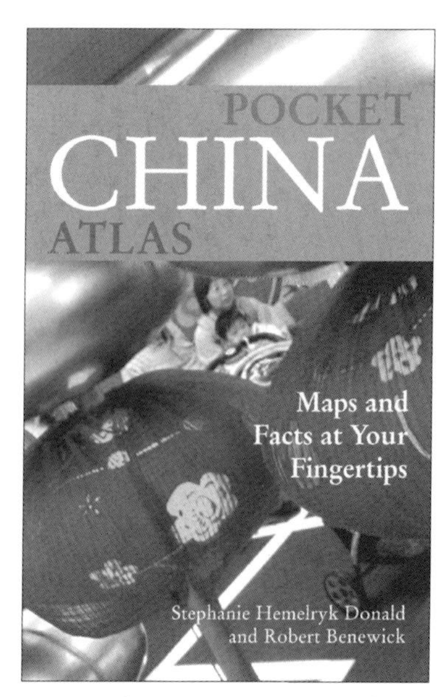

The companion mini edition
for China watchers worldwide.

In the same series:

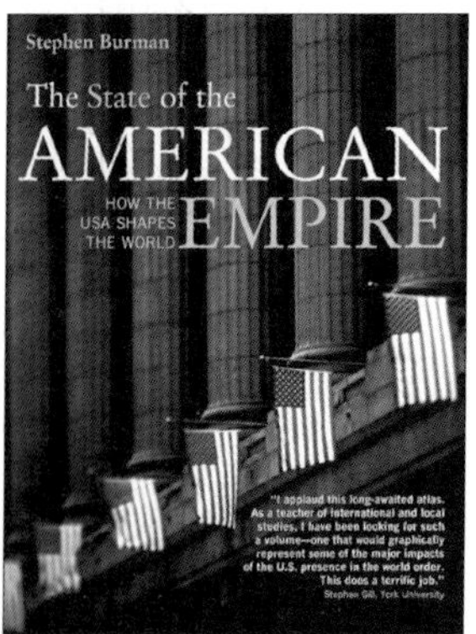

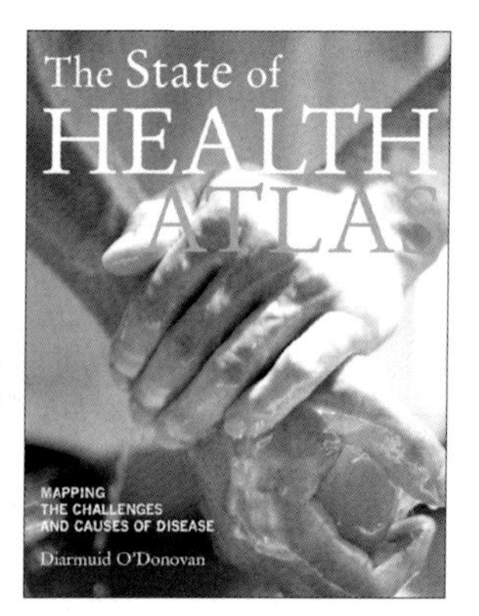

THE STATE OF **CHINA** ATLAS

Mapping the World's
Fastest-Growing Economy

Revised and Updated

Robert Benewick
and
Stephanie Hemelryk Donald

UNIVERSITY OF CALIFORNIA PRESS

Berkeley Los Angeles

University of California Press, one of the most distinguished university presses in the United States, enriches lives around the world by advancing scholarship in the humanities, social sciences, and natural sciences. Its activities are supported by the UC Press Foundation and by philanthropic contributions from individuals and institutions. For more information, visit www.ucpress.edu.

University of California Press
Berkeley and Los Angeles, California

Copyright © Myriad Editions 2009

All rights reserved

ISBN 978-0-520-25610-1 (pbk. : alk. paper)

Produced for University of California Press by
Myriad Editions
Brighton, UK
www.MyriadEditions.com

Edited and co-ordinated by Jannet King and Candida Lacey
Designed by Isabelle Lewis and Corinne Pearlman
Maps and graphics created by Isabelle Lewis

Printed on paper produced from sustainable sources.
Printed and bound in Hong Kong through Lion Production
under the supervision of Bob Cassels, The Hanway Press, London.

16 15 14 13 12 11 10 09
10 9 8 7 6 5 4 3 2 1

This book is sold subject to the condition that it shall not by way of trade or otherwise, be lent, re-sold, hired out, or otherwise circulated without the publisher's prior consent in any form of binding or cover other than that in which it is published and without a similar condition including this condition being imposed on the subsequent purchaser.

CONTENTS 中

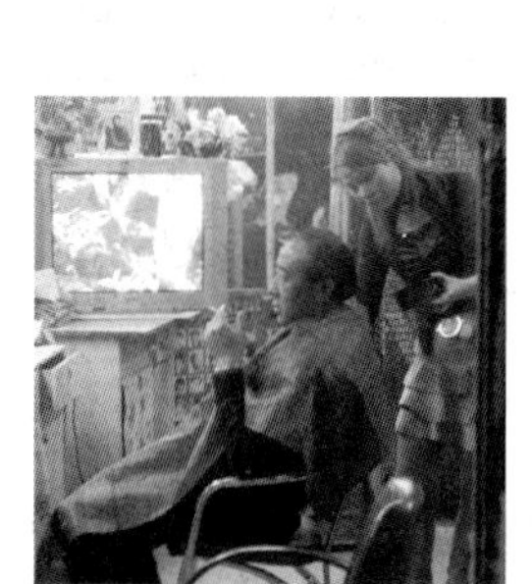

FOREWORD

THROUGH THE BEIJING SUMMER OLYMPICS of 2008 the Chinese government presented an image of a nation deeply linked to its past and increasingly confident and engaged in the world. The picture was one of power, orderly growth and development. Yet the same year also saw a major protest in Tibet, devastating natural disasters and consistent protests as the Olympic torch made its way around the world. How do we make sense of this complex society that is undergoing the kinds of changes that many of us can barely imagine? Reforms have impacted on every aspect of economy, society and state and have produced new winners and losers.

A great place to start unraveling this fascinating puzzle is the latest edition of *The State of China Atlas*. All our futures are bound up with what is happening in China, and this book is a welcome resource for anyone seeking to understand the momentous changes taking place. As the saying goes, a picture speaks a thousand words, and this is certainly true with this wonderful volume. The maps are well chosen and provide all the information that is needed by anyone wanting to get to grips with China. Even for those who have studied China for many years, the maps present new insights. For students, the Atlas is a dynamic and exciting way to bring China alive. It contains so much useful information on every topic crucial to China's development and is a key reference work. The authors have even found a way to capture the Party-State in visual form, no mean achievement!

What is revealed in these pages is a China where multiple realities operate beneath the facade of a unitary nation-state. Not only does the terrain range from the huge oceans to the east, the massive plateau to the west and the surrounding mountains that have helped China retain a certain insularity, but also the peoples of China, the climate, its industry and its agriculture show tremendous diversity. *The State of China Atlas* shows how, since 1949, the Chinese Communist Party (CCP) has placed its stamp on this varied terrain. The CCP's vision of a modern state and its policies of industrialization have had a marked impact on the physical structure of towns and countryside as well as on people's lives. On taking power, the CCP laid out a vision of the future that was inspired by the Soviet Union. A modern China would be one that was urban and industrial, with production socialized. The private sphere was to be destroyed and the rural sector was to be placed in service of the industrial push.

This produced an ugly urban landscape, and even historic cities such as Beijing had much of the grace and charm ripped out them. The industrial push made smokestack factories a familiar part of many cities, with little or no idea of zoning and protection of green areas. The countryside was also transformed, culminating in the commune movement in the late-1950s. Campaigns to boost grain production led to mountain slopes being cleared of trees, and good grazing land being plowed under. Rural industrialization led to more forests being ripped up to produce steel in "backyard furnaces", much of which was useless.

Despite such attempts to produce a dull conformity, China remained full of contradictions and variety, and these have been allowed to blossom again since economic reforms were promoted with such zeal from the late-1970s on. As the Atlas shows, these reforms have touched on every aspect of Chinese life, changing the appearance of both rural and urban China, while binding the two closer together than in the Mao years. Cities are less homogeneous than before, and the drab Stalinesque town centers have been transformed with the rise of gleaming, glass-fronted skyscrapers housing luxury offices, shopping malls and the ubiquitous McDonald's. These buildings, and designer brands such as Gucci, are the new symbols of modernization, and much of the old architecture that survived the Maoist blitz has been bulldozed out of the way. Communities have been broken up and scattered in the name of modernity. Much of the new building that is not commercial is to reify state and party power with many, new, gleaming, marble-decked buildings constructed to house the local party, government and judicial organs of the state.

Beneath the high rises, the Chinese streets are home to a much more diverse life. The markets, restaurants and discos are signs of the new entrepreneurship, or of official organizations moonlighting to make a bit of extra money. The restaurants are filled with the beneficiaries of reform: the private entrepreneurs, those involved in the new economy, the managerial elites, the politically well connected and the foreigners. There are also the millions of migrants who have poured into the cities from the countryside to build the new urban "nirvanas". They staff the construction sites, work as waiters, shop assistants, masseuses, and in the less acceptable areas of vice and prostitution. Yet not all have been blessed by this tremendous boom and economic growth. There are the new urban poor who have been laid-off from the old state-owned factories or who have no children to look after them in old age. And, as the Atlas shows, the cost of economic growth is higher levels of traffic congestion, air pollution and water contamination.

The countryside has also changed, with the collectives broken up and farming responsibility placed back with the households. This has allowed more diversification in agricultural production and has permitted millions to leave the land to find more remunerative work in the small township factories, the construction sites of big cities, or in the joint venture factories of South China. Those who move are the young, the fit and the adventurous. Those who remain to tend to the farm and household chores are the elderly, married women, the children, and the sick. The leadership under President Hu Jintao and Premier Wen Jiabao has launched major programs to try to deal with regional inequalities and to improve the living standards and access to basic services for those who have not benefitted so well from reforms.

These changes not only impact on China but have worldwide ramifications. China is now the largest recipient of foreign direct investment in the world, most multinational corporations have a China strategy and many countries are trying to align their own production strategies to meet China's development. China's economic trajectory, with its incumbent energy increases, is altering global markets for natural resources, and prices will be increasingly determined by projections of China's needs. The financial crisis of 2008 showed clearly how China's fate is intertwined with that of global economic health, and there will be consequences that move beyond the purely economic. Already, Japan is badly affected by industrial emissions from China, and the country is a major producer of greenhouse gases. Decisions made in China affect other nations in unexpected ways. For example, the ban by the Central government on logging in southwest China is eminently sensible, but has not stopped China's desire for raw wood for its domestic and export markets. This is leading to an increase in logging not only in surrounding countries such as Laos, but even in those as far away as Brazil.

The State of China Atlas provides a good starting point for trying to unravel the consequences of these changes. It is not only informative but also fun to read and look at. It is highly recommended for all those interested in the momentous changes taking place in China.

Tony Saich
Daewoo Professor of International Affairs
Harvard Kennedy School, November 2008

INTRODUCTION 中

THE 2008 BEIJING OLYMPICS REPRESENTED an important statement by an emerging regional and world power. China is culturally, economically, linguistically and politically relevant to the global community. It is a huge, complex and contradictory geo-political entity that has taken its place in our collective consciousness on its own terms. Nonetheless, China still remains mysterious to many people. Media spectacle, commerce and extended trade relations do not entirely counteract the unknowability of profound difference.

So how is China known in the contemporary world? Whilst, in the international imagination, China is bound up in extravagant symbols of development and capital, its minority peoples and most of its provinces are hardly known. Most people think of the "centralizing kingdom" (*zhongguo*), as it has been conveyed through classical art, Tang poetry, revolutionary meetings and the killings in Tiananmen protests of 1989, through to sparkling business districts in Shanghai, the spectacular historical epics of films by Zhang Yimou, and the woeful faces of Sichuanese survivors and their rescuers during the earthquake of May 2008. In media reports, China is either a friend or a foe. It is power or money, suppression or great courage. Sometimes it is a flood, the earthquake, demonstrations, or sporting valor.

China remains seemingly unknowable, because – in getting to know China – the West must recognize the limits of its assumptions, and that challenge is too hard in our own state of chronic transition and global discomfort. China is intrinsically bound up with the world's future, but its powerful national sentiments will mean that this mutual future has to be negotiated, not presumed. This is the great value in understanding the world through the state of China.

Almost every day we read that China is among the top trading nations; that its economy is one of the world's largest; and that it is one of the nations attracting the most direct foreign investment. As if this is not enough, China has become one of those powerful national economies with reasonable cash reserves, it is the government that could decide to de-invest from the USA and watch the leader of the free world go down in a whirl of debt.

There is much to trumpet, even to celebrate, and we can marvel at China's successes. Even on an ideological level, leaders of western nations derive a certain satisfaction since many of China's economic achievements can be credited to the market-led reforms that began in 1978. This may be a dominant perspective but it is not the only one. As is the case for every nation the reality is more complex. There is no doubt that most citizens in China are better off than they ever have been. Many are richer than before. An alternative perspective, however, takes into account that although there have been impressive inroads into poverty alleviation many millions remain desperately poor; a new entrepreneurial middle class, and along with that an aspirational working class, is emerging, but the income gap between each socio-economic segment is widening. A welfare system is being developed, yet healthcare remains beyond the reach of most citizens; the ageing population will soon become the largest in the world, and they will need care and support. Meanwhile, new graduates with hard-earned college degrees are scrambling to find work.

These are examples of the contradictions that confront China's elitist and insulated leadership. They are problems familiar to other nations, but they are exacerbated by the sheer scale of China's population, and by the spatial challenges and financial disparities of the country as a whole. China's population size can be seen as a great resource in a globalizing economy, providing a flourishing consumer market and a bottomless pool of cheap labor to exploit. It is also a source of mounting dissatisfaction, unrest and conflict, and so it is no wonder that China's authoritarian Party-State places political and social stability, alongside the market-led economy, as its main priorities in seeking to establish and maintain an "harmonious society". Human-rights abuses continue, despite and sometimes because of international pressure. In 2008, many ordinary Chinese citizens (overseas and in China itself) demonstrated across the world against what they regarded as international interference in Chinese affairs, and China-bashing.

Most threatening to the very fabric of the Party-State is the rampant corruption. The attempt to bring corruption under control is one of a number of reforms to the political system. Another example is the introduction of participatory, if not democratic, practices at the grassroots and basic levels of government. Reforming the Chinese Communist Party to grant more influence to the membership is also on the agenda. Whether the pace of these reforms is enough to meet the challenges to social justice is an open question.

Economic power grants China considerable leverage in international relations, but it can also be problematic. The big if-and-when question within the region and the wider international system is whether there will be a struggle for dominance between the USA and China – and whether that will play out through economics and trade or some more deadly means. The current atmosphere of perpetual war, since the war on terror was declared in 2001, is frightening and would be more so if China too were pulled into the morass. The status of Taiwan and its mainland liaison activities is a constant source of concern, especially for those who see Taiwan's democracy as a boon in the region.

Meanwhile, statisticians and demographers in China collect large volumes of statistical information, through which they might measure the inequities across provincial and regional boundaries, and with which social scientists can interpret the state of China on the ground. It may be unfashionable but perhaps these public servants are more heroic, and certainly more functional, in determining what needs to be done in China now, and for whom. As we point out in Part Seven of this book, there are no perfect statistics, and even the collection of data is subject to political controls, both in China and worldwide. Statistics alone do not explain why one city will thrive under WTO regulations, whilst another will ignore them because they threaten local political elites, or because local businesses need local subsidy to survive and protect employment. One city will put resources and imagination into branding itself in the international imagination, whilst another will miss (or deliberately ignore) the point of tourism and "destination management".

There are maps in this book, then, in which to give a truly accurate picture of the state of China, we would need to provide detailed regional specifications, county- and township-level case studies, and a lot of historical background. Where we cannot give this detail we have suggested readings from works of current China specialists, in economy, culture, the social sciences and political history, and we really hope that readers are inspired to follow up these suggestions. There are many resources on the internet, the best of which we have tried to include in the commentaries at the back of this book. Financial pages of national newspapers are also good sources for seeing certain aspects of new China unfold before our eyes. Every deal, every bankruptcy, every corporate decision will affect someone, possibly many thousands of people, in contemporary China and beyond.

As with all books, but especially one such as this, which requires a range of knowledge and expertise, the authors are indebted to the wisdom of others. The support (in time and space) from the Asian Studies and Chinese Departments and Media and Communications Department at the University of Sydney, and a fellowship at the Department of Film at Kings College London have been invaluable. The ongoing work of colleagues in Australia and overseas: Mayfair Mei-hui Yang, Ying Zhu, David SG Goodman, Harriet Evans, Michael Keane, Luigi Tomba, David Kelly, Louise Edwards, Elaine Jeffreys, is always inspirational. We also thank the Australian Research Council for its support for the Middle Class Taste project (with Zheng Yi), which has informed the arguments of this book.

We are grateful to Marc Blecher, Cynthia Enloe and Professor Hua Qingzhao, who engaged us about particular approaches taken in the atlas. The resulting interpretations are those of the authors. Sarah Cook, Rosemary Foot, Jude Howell, Jie Dao, Leicia Petersen, Peng Zou, Tina Schilbach, Ming Liang, Norman Stockman, Paul Wingrove and the truly wonderful Philippa Kelly, have made important contributions to the atlas and provided much-needed advice and encouragement for the whole project, while the University of Sussex was more than generous with the provision of facilities.

We are pleased to have been able to select photographs for pages 24, 34, 62, 74 and 88 from Beijing-based photographer Ben McMillan. In addition, we are grateful to the following for permission to reproduce their photographs: 12 and 20 Christopher Herwig; 22 (top) John Sigler/iStockphoto; 22 (bottom) Adrian H Hearn; 23 (top) Adrian Beesley/iStockphoto; 22 (bottom) Anthony Brown/iStockphoto; 22 (right) Matthew Spriggs; 52 Mark Henley / Panos Pictures; 104 DSG Goodman.

This atlas is a collective project, and our co-workers at Myriad Editions, Candida Lacey, Corinne Pearlman, Isabelle Lewis and, most of all, Jannet King deserve to have their names on the cover and title page with us. We thank them with great affection.

Memories of Anne Benewick (1937–1998), the co-founder of Myriad, inspired us. China held a special place among the Myriad Atlases, for China was special for Anne. When in Beijing she played a mean game of table tennis, to the surprise and delight of her opponents and on-lookers. At the People's University she made *jiaozis* with faculty and students, cycled on her second-hand Flying Pigeon and searched relentlessly for the perfect peony.

Robert Benewick
Brighton, UK

Stephanie Hemelryk Donald
Sydney, Australia and London, UK

Trade between China and Africa increased 67% in the first half of 2008.

Part One

CHINA IN THE WORLD

CHINA'S REMARKABLE ECONOMIC PERFORMANCE catapulted the country on to the world stage in the late 1990s and has continued to support the country's claim for world status ever since. In the first half of 2008, the year of the infamous "credit crunch", China was responsible for one-third of global GDP growth, and despite a previous high of 12 percent growth per annum, even a global downturn is likely to slow China only to a respectable 8 percent. Given the undoubted fragility of nations whose economy is fuelled by debt, and whose systems are imperiled by banks that gamble on the future in order to create virtual money for dividends in the present, China's dual role as the world's biggest saver and manufacturer very probably stands it in good stead for the next wave of capitalism.

Nor is there any doubt about China's ambitions as a world power. Not only is it a major manufacturer and, more recently, product and brand developer, but there have been significant developments in its profile in international relations. Its newest venture is in Africa, where it has invested billions in infrastructure designed to extract the continent's resources. China's involvement in Africa is experienced by the West as a challenge to its own colonial inheritance of trading pre-eminence and geo-political strategic strength. The modernization of the Chinese armed forces and the extent of China's investment in diasporic interests worldwide also contribute to a striking increase in China's national strength.

Despite some areas of tension, especially in the western border provinces of Tibet and Xinjiang, China presents a sense of national wholeness. Both people and government, at least in the main, appear confident and proud of what has been achieved to date, and what will be done in the future, if the economy and stability of the country is managed well. That is the crux perhaps of China's success, but also of its vexed relationship with western democracies, especially its largest debtor, the USA.

China has undoubtedly transformed its economy and governance structures in a relatively short space of time. Since the accession of Deng Xiaoping in 1978, the leadership has been most interested in legitimizing the Communist Party and the State, moving away from a command economy towards a semi-market system. There is little doubt that many people in China are materially better off thanks to this policy, although the gaps in income and opportunity between the new rich, the working middle classes, and the poor, and migrant underclasses increase exponentially with every leap in growth.

The second priority has been to ensure stability. The older generation of leaders had direct experience of the Cultural Revolution and its impacts on their generation were deeply scarring. One result has been the maintenance of a rigid authoritarian system of rule, with power located in Beijing, albeit including local governance reforms and some dispersal of influence amongst provincial governments. The two priorities – economic growth and strong leadership – are conjoined in China's current approach to the world: growth and investment is desirable, but anything that undermines sovereignty or stability will be dealt with extremely firmly. This suits investors, but leads to transnational disputes over emotionally charged issues such as Tibet. On the other hand, most nations in the West realize that without China's leadership and interventions in North Korea, the whole Pacific region would be significantly less stable.

WHEN AN ARROW IS ON A STRING IT MUST GO

By 2006, China, in conjunction with Hong Kong, was the second-largest trading nation in the world – up from 15th in 1990.

Despite importing billions of dollars of raw materials to feed its hungry industrial manufacturing sectors, China maintains an overall trade surplus. This is in stark contrast to the USA which, despite being the largest trading nation in the world, is in deficit.

China's economy has been growing at an extraordinary rate, and trade plays a vital role in the country's economic liberalization and modernization. Between 2003 and 2007, exports – almost all of which are manufactured commodities – nearly trebled.

Although European traders have progressively increased their share of China's markets, the overwhelming bulk of China's trade, and its fastest-growing partnerships, are with its Asian neighbors. However, China's development strategies in Africa, Latin America and the Oceanic island nations suggest that these regions are increasingly providing China with markets as well as raw materials.

▸▸ *see also page 105*

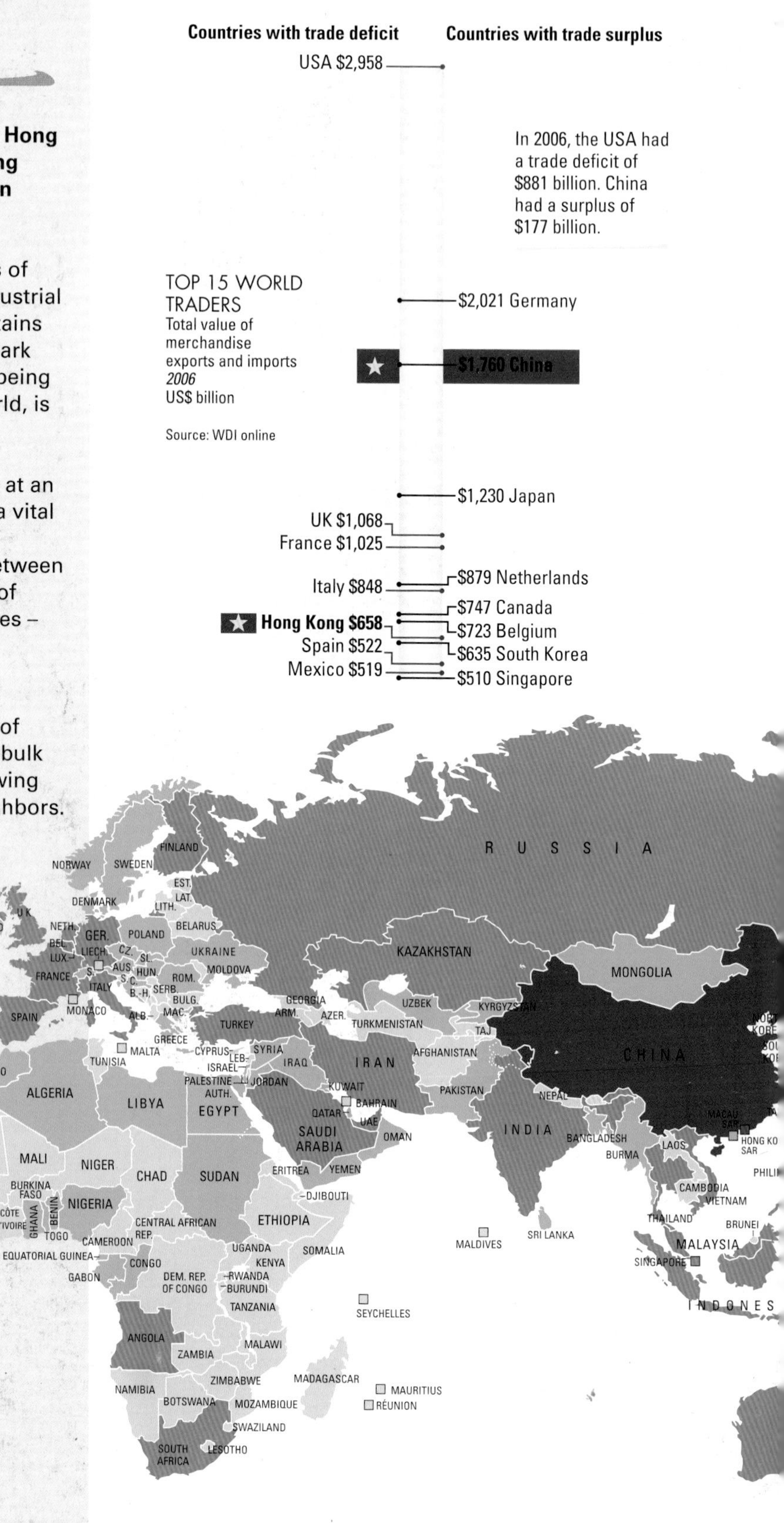

 Copyright © Myriad Editions

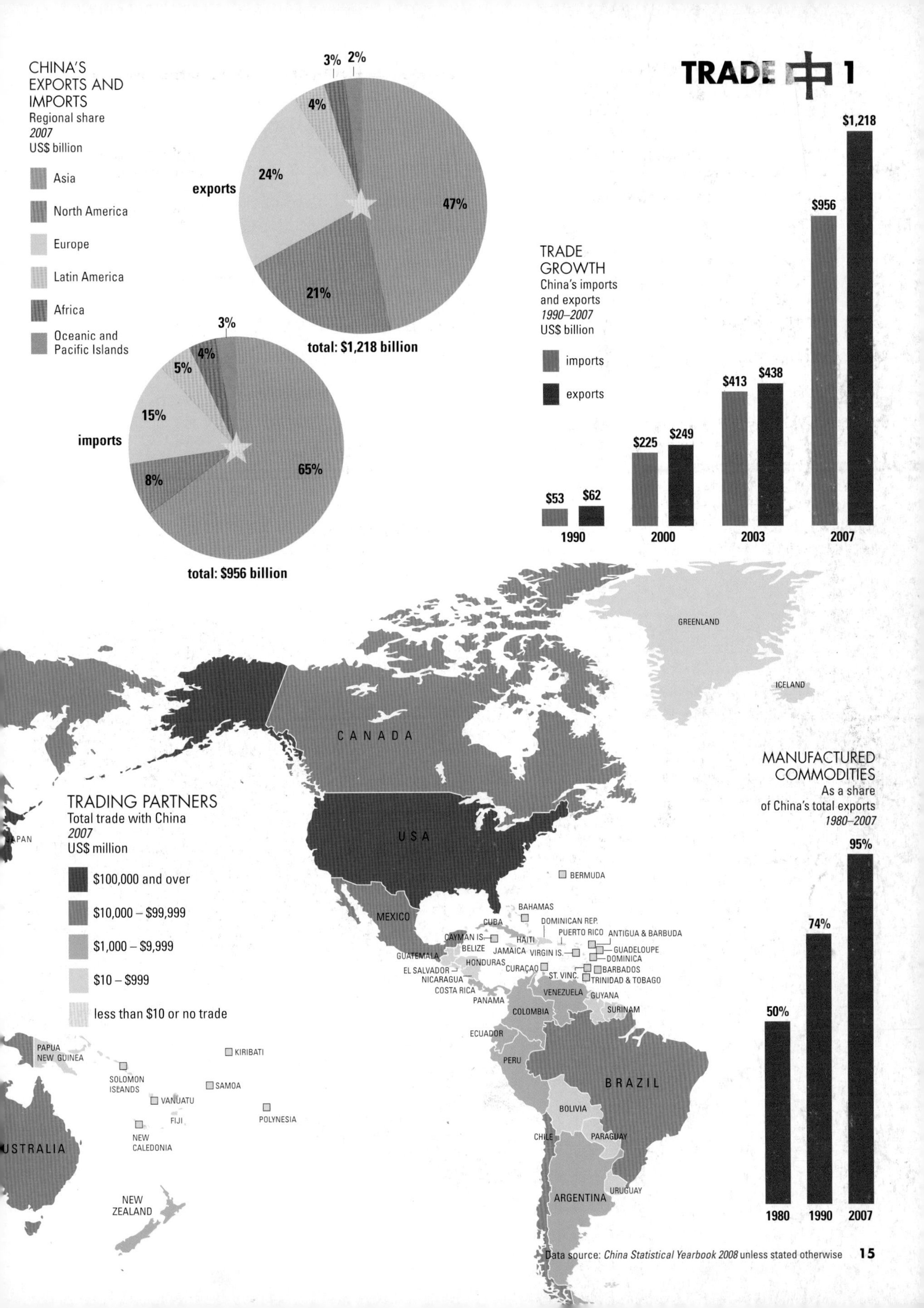

TRADE 中 1
CHINA'S EXPORTS AND IMPORTS
Regional share
2007
US$ billion
Asia
North America
Europe
Latin America
Africa
Oceanic and Pacific Islands
exports
47%
21%
24%
4%
3%
2%
total: $1,218 billion
imports
65%
8%
15%
5%
4%
3%
total: $956 billion
TRADE GROWTH
China's imports and exports
1990–2007
US$ billion
imports
exports
$53
$62
1990
$225
$249
2000
$413
$438
2003
$956
$1,218
2007
TRADING PARTNERS
Total trade with China
2007
US$ million
$100,000 and over
$10,000 – $99,999
$1,000 – $9,999
$10 – $999
less than $10 or no trade
GREENLAND
ICELAND
CANADA
USA
MEXICO
BERMUDA
BAHAMAS
CUBA
DOMINICAN REP.
PUERTO RICO
ANTIGUA & BARBUDA
CAYMAN IS.
HAITI
BELIZE
JAMAICA
VIRGIN IS.
GUADELOUPE
DOMINICA
GUATEMALA
HONDURAS
CURAÇAO
BARBADOS
EL SALVADOR
NICARAGUA
ST. VINC.
TRINIDAD & TOBAGO
COSTA RICA
PANAMA
VENEZUELA
GUYANA
SURINAM
COLOMBIA
ECUADOR
PERU
BRAZIL
BOLIVIA
PARAGUAY
CHILE
ARGENTINA
URUGUAY
PAPUA NEW GUINEA
KIRIBATI
SOLOMON ISLANDS
SAMOA
VANUATU
FIJI
POLYNESIA
NEW CALEDONIA
USTRALIA
NEW ZEALAND
JAPAN
MANUFACTURED COMMODITIES
As a share of China's total exports
1980–2007
50%
1980
74%
1990
95%
2007
Data source: China Statistical Yearbook 2008 unless stated otherwise
15

JUMPING INTO THE SEA

Foreign direct investment in China continues to grow and is significantly greater than that in its Asian neighbors.

Overseas companies are tempted by the large pool of available labor, the 8 percent annual rate of growth over several years, and the maturing, although still youthful, stock market. China's accession to the World Trade Organization in 2001 has enabled foreign firms to enter into partnerships with Chinese companies. Manufacturing has been dominant, although the growth of the services sector has seen foreign entries in real estate, hospitality, retail and communications. The leading sources of investment are from North America, Europe and the Asia-Pacific region. Overseas Chinese, even those living in relatively small and island economies, choose to invest in China's future.

Not all entrepreneurial ventures run smoothly. The regulatory framework around quality controls is uneven in theory and practice, which can cause problems for co-partners in certain enterprises. Piracy and copyright remain a problem for some forms of investment.

China's own direct investment overseas is of huge importance to the global economy, but has political implications, not only in Africa, but in the USA and Australia.

▸▸ *see also page 105*

NET INFLOWS OF FOREIGN DIRECT INVESTMENT (FDI)
Selected countries in Asia
2006
US$ billion

Source: WDI online database

78.1 China
42.9 Hong Kong
24.2 Singapore
17.5 India
9.0 Thailand
3.6 South Korea
2.3 Phiippines, Vietnam

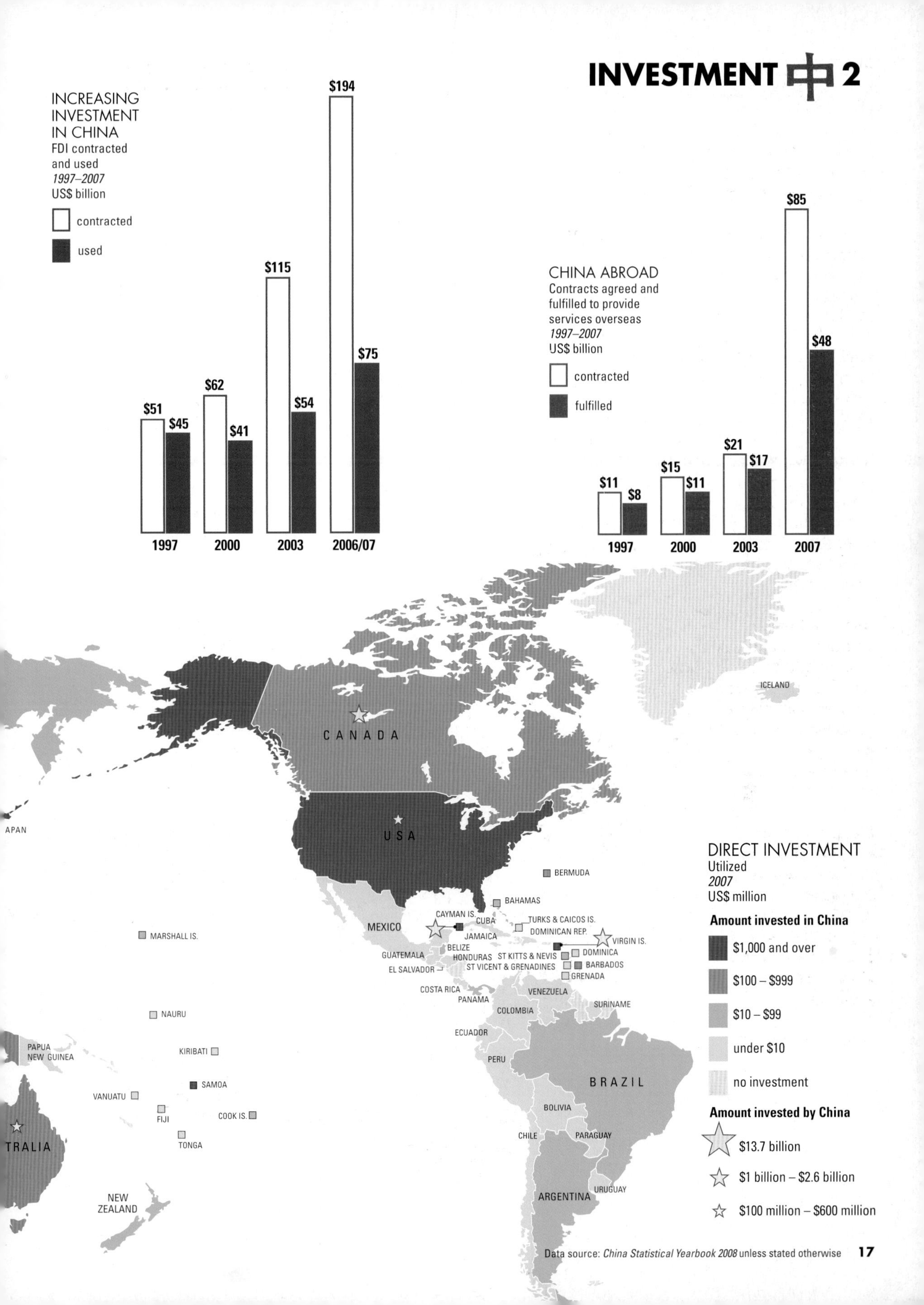

Data source: *China Statistical Yearbook 2008* unless stated otherwise

THE MASTER DOES NOT FIGHT; WERE HE TO DO SO HE WOULD WIN

China's regular armed forces number 2.1 million, and comprise 9 percent of the world's total. In addition, there is a reserve force of 800,000, and a military police force of 1.5 million.

An army of this size is impressive on paper, and reasonably cheap to run, but in modern warfare it is high-tech weaponry that counts, and China's military expenditure is a long way behind that of the USA and the total for the rest of NATO. China does have a nuclear capacity, however. While minuscule in comparison with that of Russia or the USA, it is significant in terms of China's military power within Asia and South-East Asia.

China's unannounced testing of an anti-satellite missile on an ageing weather satellite in January 2007 is also significant. Military specialists noted that it involved the successful interception of an object travelling on a similar trajectory, and at a comparable speed to that of an inter-continental ballistic missile.

China has become increasingly keen to play a part on the world stage, with around 2,000 troops, engineers and medical staff involved in peacekeeping missions around the world during 2008, including UN and African Union missions in Sudan. It has also, however, been criticized for its sale of arms to developing countries, in particular to the Sudanese government.

▸▸ *see also page 106*

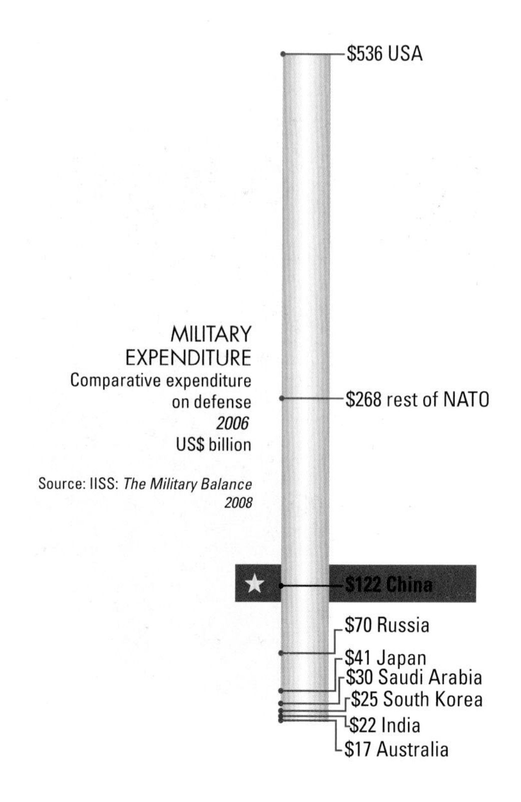

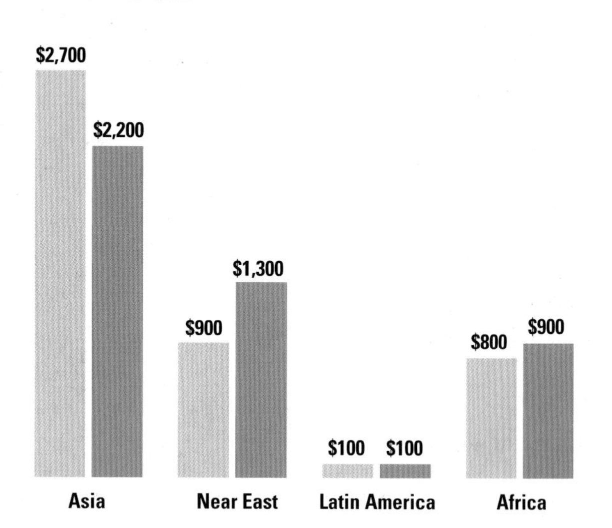

 Copyright © Myriad Editions

MILITARY POWER 中 3

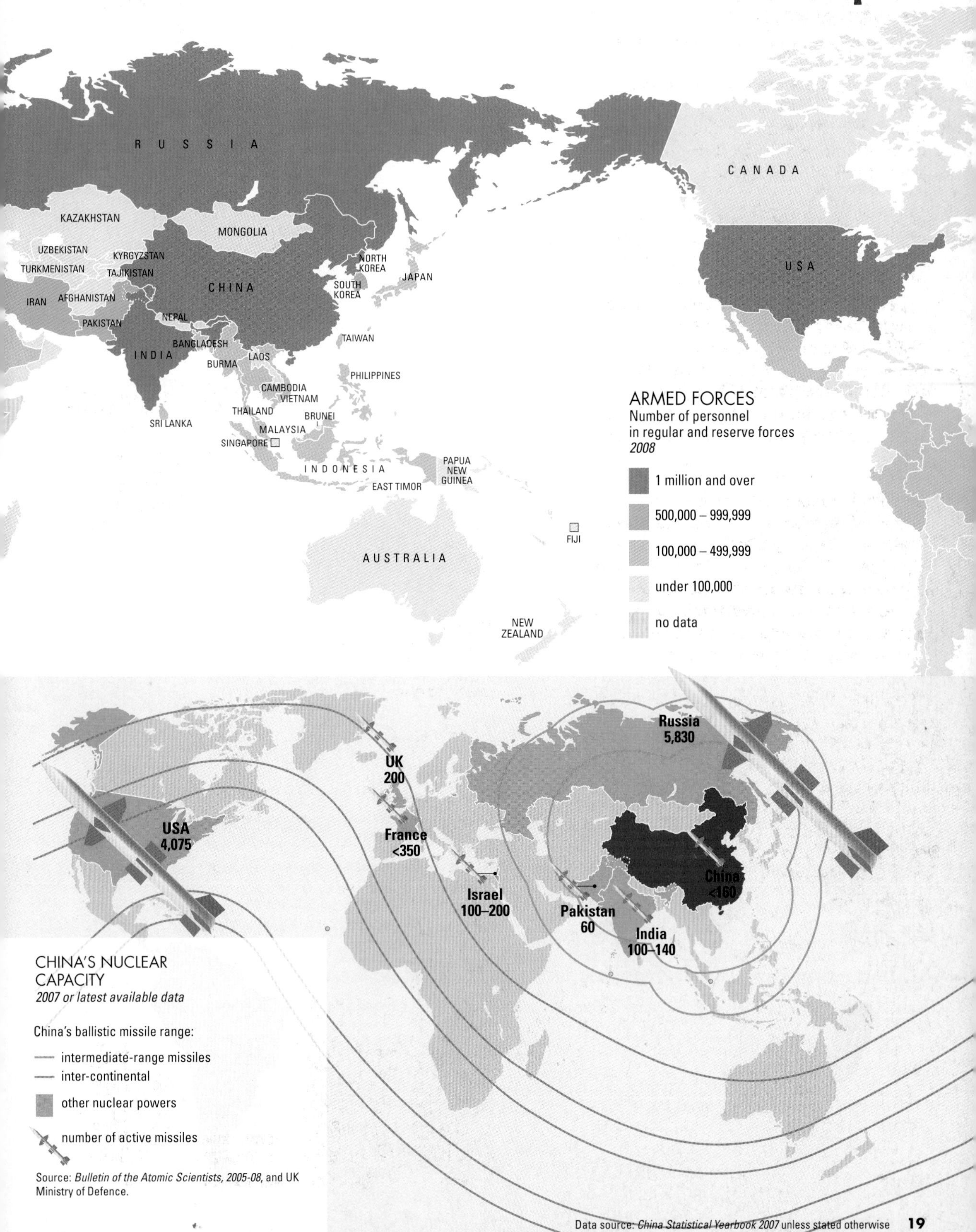

THE FIRST UNDER HEAVEN

China's government manages the country's international relations both through its alliances and negotiated positions with overseas powers, and by taking into account its people's far-reaching ambitions and passionate nationalism.

As China's overall position in the Asia–Pacific region strengthens, and it develops its economic interests in Africa and Latin America, so its relations with former world powers and the fragile economic "giants" of the USA and Europe become more volatile.

The USA, in particular, retains deep suspicion of Chinese intentions, and the Chinese people are also ambivalent about the USA and its attitudes to them. Originally seen by some as a source of freedom, or simply the fount of capitalist leadership, the USA is now also recognized as a competitor that does not necessarily respect China's government or domestic policy, and which actively fears its expansion in the Pacific region. Taiwan remains a thorn in the side of smooth relations between these two major powers.

The other major issue of contention is human rights. The Chinese government attributes the discourse of human rights to Western liberalism, and individualism, and anti-Chinese aggression, whereas most in the developed West understand the concept as the cornerstone of internationals standards and democratic process. This is a bugbear of international relations and mutual comprehension for both parties.

▸▸ *see also page 106*

CHINA'S CHANGING INTERNATIONAL RELATIONS
1989–2008

- action detrimental to China
- action beneficial to China

China–European Union (EU)

- **1995-2008** Annual Human Rights Dialogue.
- **1998–2008** Annual Summit Meetings.
- **2003** China releases first policy paper on the EU.
- **2005** Joint Declaration on Climate Change.
- **2005** Co-operate on Galileo satellite navigation.
- **2007** High-level economic and trade dialogue established.
- **2007** Agree to enhance co-operation on climate change.
- **2007** China among 16 ASEAN countries leaders in trade with the EU.
- **2008** European Parliament awards Sakharov Prize for Freedom of Thought to Chinese dissident Hu Jia.
- **2008** China–EU School of Law opens in Beijing.

China–Africa

- **2006** China-African Summit held in Beijing. Business deals, loans and credits agreed.
- **2006** One of China's largest oil and gas producers buys stake in Nigerian offshore oil and gas fields.
- **2007** Hu Jintao visits eight African countries to boost trade and investment.

- **2007** China opposes sanctions on Darfur.
- **2008** China to double assistance to Africa by 2009.

 Copyright © Myriad Editions

China–Russia

- **2004** Strategic energy deal agreed.
- **2005** Joint military exercises are held.
- **2006** Energy co-operation extended.
- **2007** Commercial deals worth $4 billion are agreed.
- **2007** Shanghai Co-operation Organization holds joint military exercises.
- **2008** New Russian President Dimitry Medvedev visits Kazakhstan and China.
- **2008** China-Russia Border Treaty signed.

China–USA

- **2000** Clinton signs law giving US normal trade relations with China.
- **2001** Collision between Chinese jetfighter and US spyplane.
- **2001** China supportive of US actions in Afghanistan.
- **2002** President George W Bush visits Beijing; General Secretary Hu Jintao visits USA. USA adds East Turkestan Islamic Movement to its list of terrorist organizations. China votes for UN ultimatum to Iraq.
- **2004** US Secretary of State Colin Powell visits Beijing and strengthens US commitment to One-China Policy.
- **2004** China and USA hold first strategic dialogue. These become increasingly important.
- **2004** Bush visits China. China puts forward five proposals to better China–US relations.
- **2005** US Defence Secretary Donald Rumsfeld visits China, marking a substantial improvement in the military exchanges between the two countries.
- **2006** Hu Jintao visits the USA and holds further meetings with President Bush at the G8 Summit and at the APEC Economic Leaders Meeting.
- **2006** First China–US Strategic Economic Dialogue held.
- **2006** Rumsfeld characterizes US policy towards China as "congagement", moving from engagement towards containment/confrontation.

China–Asia Pacific

- **2001** China and ASEAN agree to establish a free-trade zone by 2010.
- **2002** China and ASEAN sign code of conduct for the South China Sea.
- **2002** First Chinese state visit to India for over a decade.
- **2003** China and India reach de facto agreement on the status of Tibet and Sillim.
- **2004** China signs trade agreement with 10 South-East Asian nations.
- **2005** First meeting between Taiwan Nationalist and Chinese Communist Party leaders since 1949.
- **2005** Anti-Japanese protests over alleged textbook glossing over Japan's World War II record.
- **2007** Prime Minister Wen Jinbao addresses Japan's parliament.
- **2008** President Hu Jintao is first Chinese head of state to visit Japan since 1998.
- **2008** China and Japan agree to jointly develop a gas field in East China Sea.
- **2008** China and Taiwan agree to set up representative offices in each other's territory. The first direct flight takes place and a deal is signed to expand air and shipping links.
- **2008** China continues to broker six-party talks on the denuclearization of North Korea.

China–Central and South America

- **1999** China sets up two intelligence stations in Cuba.
- **2004** President Hu Jintao visits Argentina, Brazil, Chile and Cuba and pledges investments of more than $100 billion in Latin America over the following decade.
- **2005** China is Latin America's third-largest trading partner.
- **2006** China fails to support Brazil for a permanent seat on the UN Security Council.
- **2006** China and the USA hold their first talks on China's role in the region.
- **2008** China provides Cuba with aid following two hurricanes.

WEARING NEW SHOES BUT FOLLOWING OLD PATHS

Chinese outward migration over the centuries has created some strong overseas communities.

The history of Chinese outward migration is complex, but the reasons why people migrated in the 18th and 19th centuries – as indentured laborers, traders, and for education or adventure – still hold true today.

Long-term Chinese residency and trade are symbolized in the Chinatowns of major cities such as London, San Francisco, Paris, Havana, and in the ordinary "Chinese" suburbs in other, less famous, places. There are Chinese communities, not only in the heavily populated countries of Europe and South-East Asia, but also in tiny Pacific island nations such as Papua New Guinea, Vanuatu, and the Solomon Islands.

In most of these communities, people of different generations of migration have vastly uneven levels of "Chinese" identity, and varying access to and knowledge of Chinese languages, of which there are several in common use. In the US census, Chinese, Japanese, Filipinos and others are categorized as "Asian-American". There are strongly felt arguments about what this means and whether or not the terminology is helpful.

The present rise of the Chinese economy has attracted return migration, and also a flow of inward investment from overseas Chinese into mainland businesses.

▶▶ *see also page 106*

CHINESE COMMUNITIES
Selected
2008

San Francisco, USA

Established in the 1850s, San Francisco's Chinatown is the largest in North America and one of the city's top tourist attractions. The Chinese population is a sizeable minority of the total. Languages spoken in the Chinese communities include English, Cantonese, Fujianese, Taishanese, and Mandarin.

Havana, Cuba

Between 1847 and 1874, 120,000 Cantonese indentured workers were introduced into Cuba to replace African slave labor in the sugar industry. By 1925 a further 30,000 Chinese had arrived, both directly from China and via San Francisco, establishing Havana's Chinatown as the most important in Latin American. Since the mid-1990s the Cuban government has attempted to reverse decades of decline by revitalizing the neighborhood, in collaboration with local Chinese ethnic associations and with support from the People's Republic of China.

Lima, Peru

Chinese laborers first arrived in Peru in the mid-19th century. Lima's Barrio Chino is home to the country's main Cantonese community and over 6,000 Chinese restaurants. Chinese migration to South America is now part of a larger managerial push, moving production of manufactured Chinese goods offshore.

CHINESE POPULATIONS WORLDWIDE
Twenty largest populations
2008

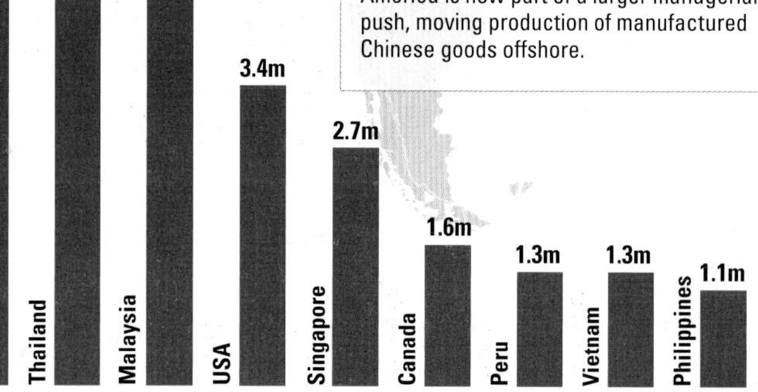

Source: Overseas Compatriot Affairs Commission, Taiwan

 Copyright © Myriad Editions

Liverpool, UK

The 7,000-strong Liverpudlian Chinese community, many of whose ancestors arrived in the late 19th century, is Europe's oldest. Liverpool has benefited from this long association with China, and is planning a "history of Liverpool" space at Shanghai's International World Expo in 2010, as well as deals that bring Liverpool's soccer culture to an eager Chinese fan base.

Russia

There are two main Chinese communities in Russia – one in Moscow, and the other in Vladivostok and nearby cities, to which Chinese immigrants are attracted by the relatively high wages on offer.

London, UK

Chinese settlement in London was originally associated with Limehouse docks. The modern Chinatown in London's Soho was established in the mid-1970s as a commercial centre that attracts investment from overseas and mainland Chinese businesses, and successfully represents the "China" brand. Its streets provide a focus for Chinese migrants seeking work and networks in the UK, as well as a familiar meeting place for international students of Chinese heritage.

Vanuatu

Chinese first arrived in the islands of Vanuatu as ships cooks and traders in the 1840s, and there are now Hakka, Mainland, Taiwanese and Cantonese Chinese living there. Both Chinese multinationals and the Chinese government have invested in the main island's infrastructure. The statue outside the Parliament in Port Vila is a gift from the Chinese government, and the work of a socialist realist sculptor.

Singapore and Malaysia

Singapore is the only country outside China where ethnic Chinese comprise the m..,·ity of the population – over , ´– although within that there are a number of language-defined communities. The Chinese population in Malaysia represe·'s only 24% of the total, and althoug ·ng-term and multi-gener...nal, it is less powerful than that in Singapore. The Chinese languages (Cantonese, Mandarin, Hokkien, Hakka, Hainan, Foochou) are described as secondary dialects to the official language: Bahasa Melayu.

Sydney, Australia

Sydney's Chinatown is the third to hold that name. The first Chinese probably arrived around the time of the First Fleet, in 1788, and a strong trading relationship has existed down the West Pacific Rim ever since. Despite the iniquitous anti-Chinese immigration policy of 1901–73, 7% of Sydney's population now defines itself as of Chinese ethnicity. Dixon Street is the focus of a large and diverse community, with restaurants offering everything from north-Chinese dumplings to seriously hot Sichuan cuisine.

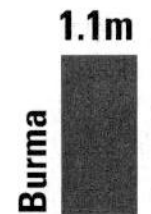

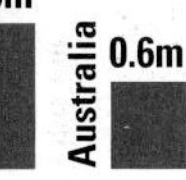

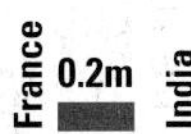

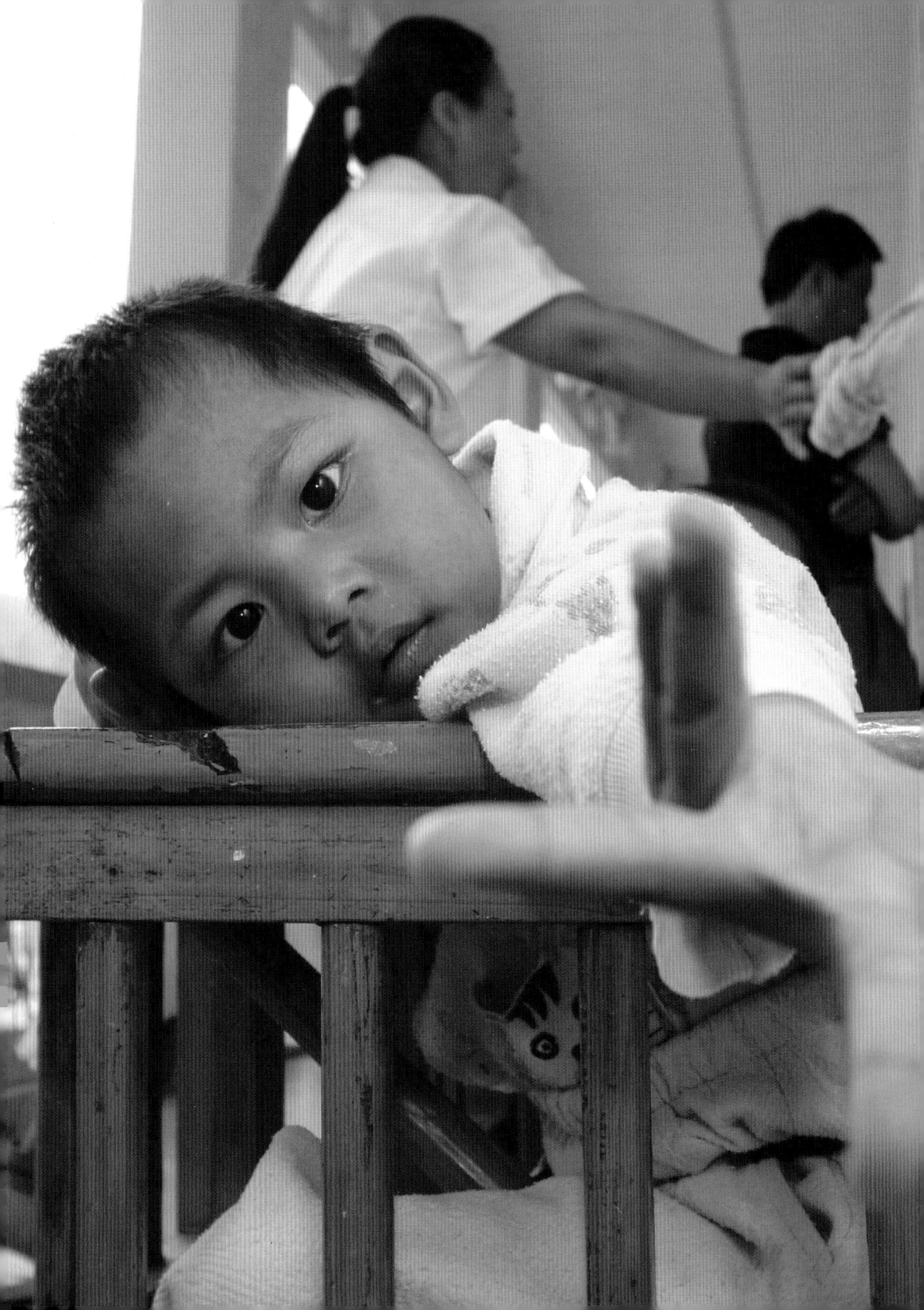

"It is easier to find Chinese-ness rooted in history than in the shared qualities among people known as Chinese."
Wang Guangwu

Part Two
PEOPLE IN CHINA

CHINA HAS A HUGE POPULATION and a vast landmass. For many years, this single fact has been the most significant characteristic of both its potential and its challenge. Yet, any understanding of China must also take into account particular sectors of the population: its minority ethnic peoples, its class divisions, and its rural and urban split – those people who dominate the policy agenda and those who are under-represented in the ideological management of the nation's future.

Research into China is usually premised on the differences between rural and urban living – an approach encouraged by the data published in the annual *China Statistical Yearbook* and in the registration system for residency in different zones. This dichotomy between the urban intelligentsia and the peasantry has a long history and although it is now being challenged by mass inward migration to the towns and cities, it is still true to say that the majority of poorer Chinese are rural, and that privilege is mainly confined to metropolitan areas.

In the past 15 to 20 years more attention has been paid to the differences between China's provinces, which are important units of experience, economy and culture. They are also relevant in central planning mechanisms, which are increasingly mapped across macro-regions. At the 2008 11th National People's Congress, Premier Wen Jiabao affirmed that the development strategies in the western provinces would be strengthened, with a focus on social support as well as industrial investment.

The Premier's National People's Congress speech, always crucial to understanding policy directions, also emphasized that rural family planning would be more tightly monitored and that disincentives to have more than one child would be relaxed. This indicates that the problem of gender preference in seeking abortions, and in the differential care of babies born in poor households, has been noted by the government.

Gender issues are high on the agenda for many organizations for a number of related reasons. Chinese girls have very uneven expectations within society, depending on their birthright. Educated and well-off families give a daughter great moral and financial support in order to further her education and future career prospects. The poor and ill-educated are more likely to give up on their daughter's prospects because of poverty, a culture of preference for male interests, or both. Where young rural women are in short supply, this leaves them vulnerable to abuse, and physical danger. Many migrate to the cities and towns for work. These girls generally end up working in factories, as maids (*baonü*), or in the service industries. Factory conditions in the south of the country are especially harsh, and there are major problems with health and safety in these hothouses of the new economy.

The Han Chinese population is in the overwhelming majority, and the term "Chinese" presupposes many Han beliefs and ways of thinking. It also, however, includes many that are originally from other ethnic groups – or that have been entirely made up or re-invented. Indeed, one can argue that "Chinese-ness" is an artificial construct that can be re-negotiated, depending on language, cultural practices and place of residence. Arguably, shared practices are as likely to be found in the border cities of Tibet, Gansu and Xinjiang, amongst people who are ethnically diverse but geographically proximate, as they are amongst Han Chinese across the nation and beyond.

YOU CANNOT WRAP A FIRE IN PAPER

Over 1.3 billion people, one-fifth of the world's population, live in China.

The sheer numbers involved affect all aspects of life. The population continues to increase – even though the rates of growth have slowed. As the population clock suggests, these numbers challenge available solutions.

China's population is unevenly distributed across its provinces. Urban areas are becoming more overcrowded as the rural population leaves the land to work in the towns and cities, especially those in the eastern region.

In 2000, China conducted the world's largest census. Despite the difficulties in taking an accurate account of such large numbers, and ensuring the cooperation of local officials, it was pronounced a great success.

▸▸ *see also page 107*

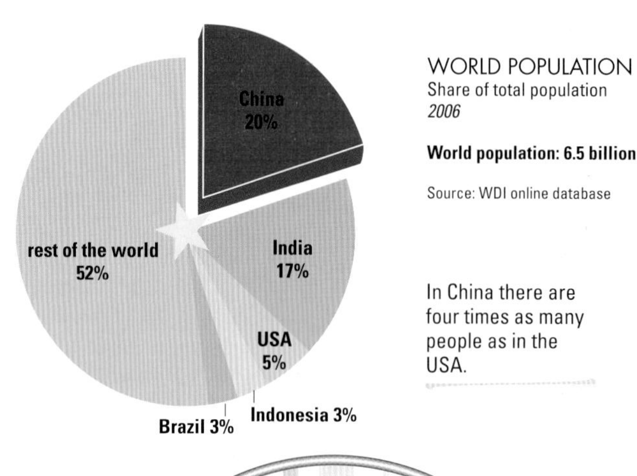

WORLD POPULATION
Share of total population
2006

World population: 6.5 billion

Source: WDI online database

In China there are four times as many people as in the USA.

POPULATION CLOCK
Number of births in China
2007

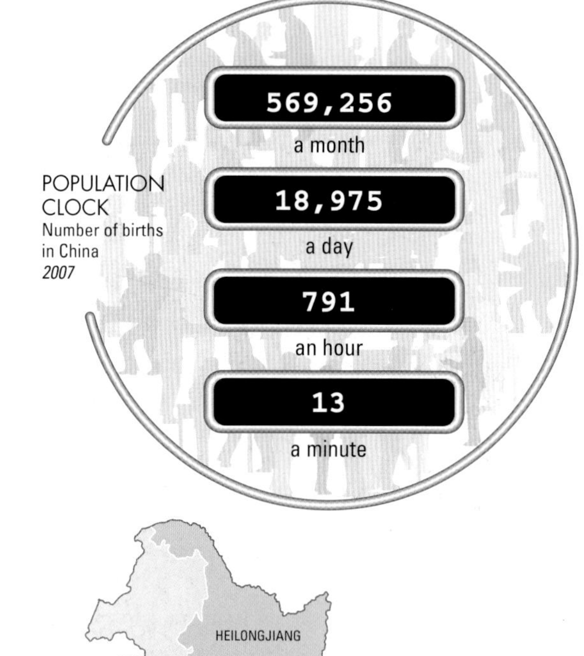

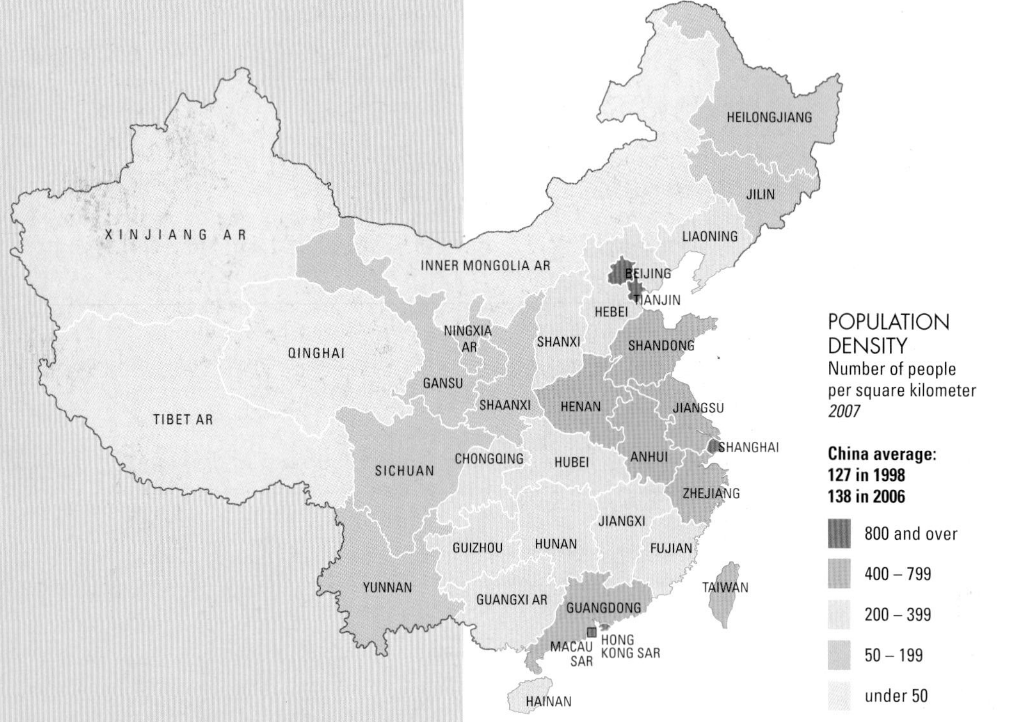

POPULATION DENSITY
Number of people per square kilometer
2007

China average:
127 in 1998
138 in 2006

- 800 and over
- 400 – 799
- 200 – 399
- 50 – 199
- under 50

 Copyright © Myriad Editions

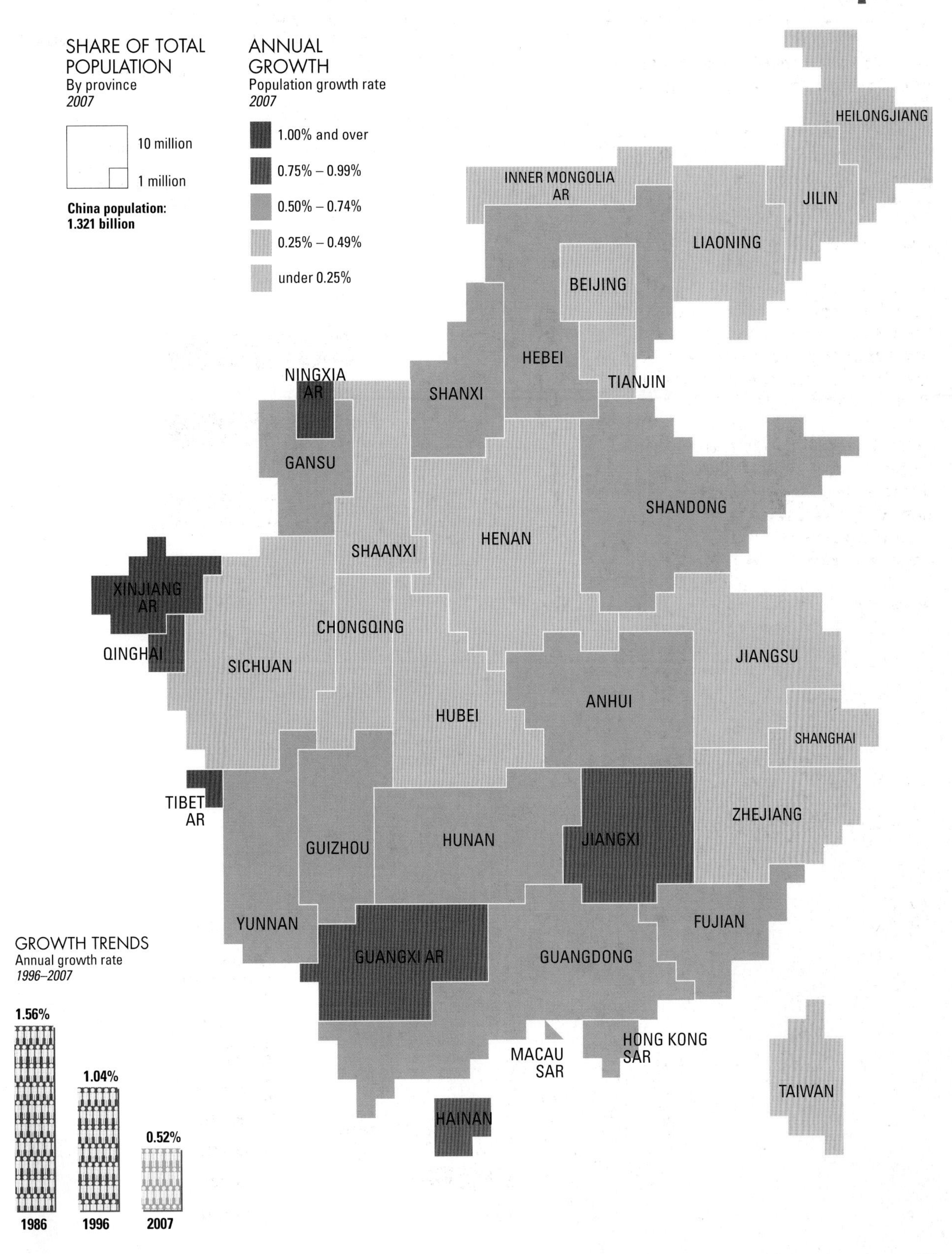

Data source: *China Statistical Yearbook 2008* unless stated otherwise

REARING A TIGER IS TO INVITE FUTURE TROUBLE

There are more boys than girls in China. This is particularly true in the rural areas, where boys are seen as more productive in agricultural work, and more valuable to aging parents.

The consequences of small- or one-child family policies has been severe for girl children in the countryside. There, stories of abandonment and neglect are common, and tales of abductions – of adolescent girls and young women – suggest a widening gender gap amongst under 30-year-olds.

Implementation of the law is left to provincial governments, who may vary it according to local conditions. Concerned about a lack of people to care for its increasingly elderly population, Shanghai has relaxed its regulations to allow some couples to have two children, and now offers incentives to daughter-only families.

Some rural women have been subject to forced sterilization in certain provinces – the same areas that now have the worst gender ratio.

▶▶ *see also page 107*

GENDER INEQUALITY
Gender-related Development Index (GDI) score
2005

The GDI is a combined measurement of life expectancy, literacy and earnings for women, compared with those for men. The higher the score, the more equal the society.

most equal
.960 Australia
.944 Spain, UK
.937 USA
.801 Russia
.798 Brazil
.776 China
.721 Indonesia
.600 India
.456 Nigeria
least equal

China is the only country where more women than men commit suicide.

Between 2001 and 2006, female and male sterilization each decreased by 3%, while condom use increased by 5%.

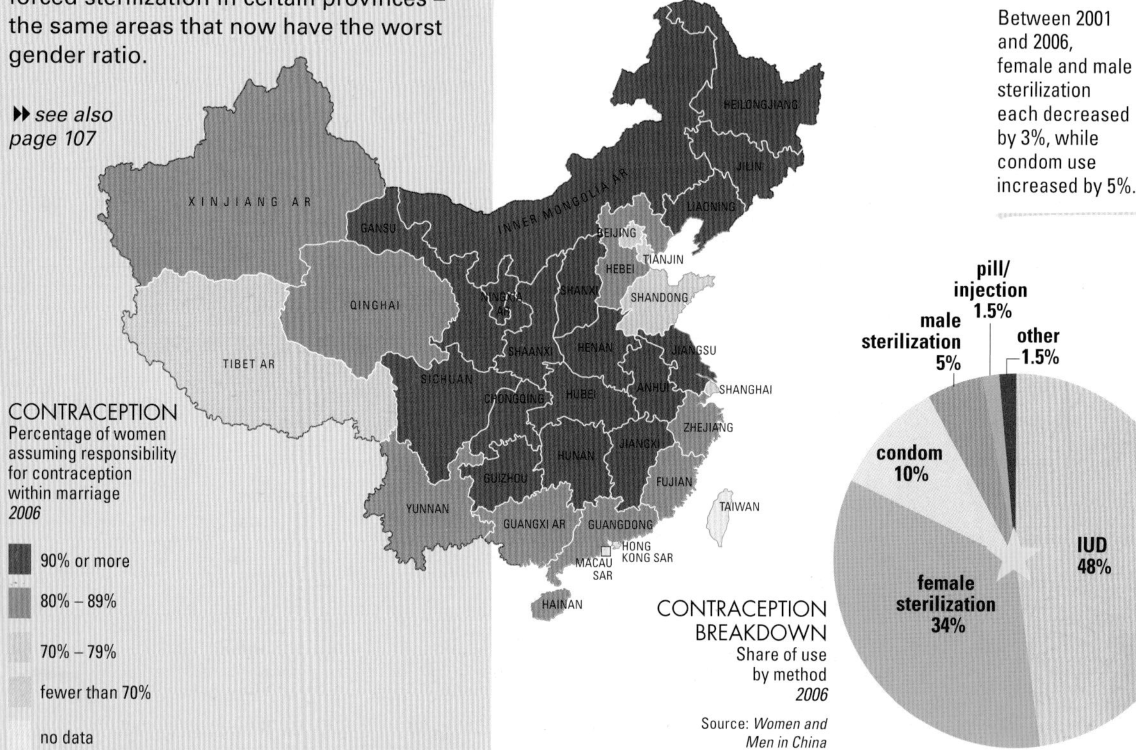

CONTRACEPTION
Percentage of women assuming responsibility for contraception within marriage
2006

CONTRACEPTION BREAKDOWN
Share of use by method
2006

Source: *Women and Men in China*

 Copyright © Myriad Editions

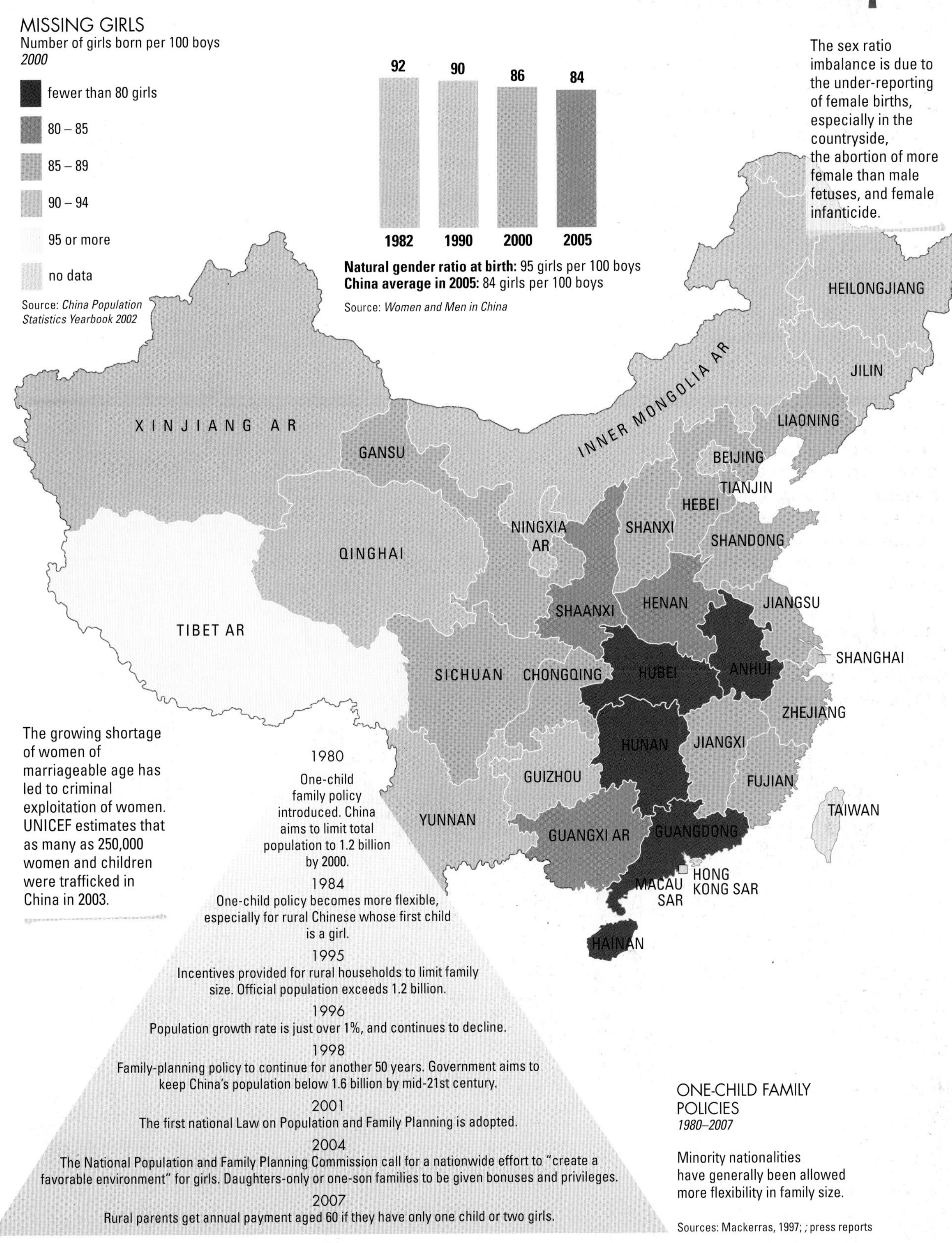
MISSING GIRLS
Number of girls born per 100 boys
2000
fewer than 80 girls
80 – 85
85 – 89
90 – 94
95 or more
no data
Source: China Population Statistics Yearbook 2002
92
90
86
84
1982
1990
2000
2005
Natural gender ratio at birth: 95 girls per 100 boys
China average in 2005: 84 girls per 100 boys
Source: Women and Men in China
The sex ratio imbalance is due to the under-reporting of female births, especially in the countryside,
the abortion of more female than male fetuses, and female infanticide.
HEILONGJIANG
JILIN
LIAONING
INNER MONGOLIA AR
XINJIANG AR
GANSU
BEIJING
TIANJIN
HEBEI
NINGXIA AR
SHANXI
SHANDONG
QINGHAI
SHAANXI
HENAN
JIANGSU
TIBET AR
SHANGHAI
SICHUAN
CHONGQING
HUBEI
ANHUI
ZHEJIANG
HUNAN
JIANGXI
GUIZHOU
FUJIAN
TAIWAN
YUNNAN
GUANGXI AR
GUANGDONG
HONG KONG SAR
MACAU SAR
HAINAN
The growing shortage of women of marriageable age has led to criminal exploitation of women. UNICEF estimates that as many as 250,000 women and children were trafficked in China in 2003.
1980
One-child family policy introduced. China aims to limit total population to 1.2 billion by 2000.
1984
One-child policy becomes more flexible, especially for rural Chinese whose first child is a girl.
1995
Incentives provided for rural households to limit family size. Official population exceeds 1.2 billion.
1996
Population growth rate is just over 1%, and continues to decline.
1998
Family-planning policy to continue for another 50 years. Government aims to keep China's population below 1.6 billion by mid-21st century.
2001
The first national Law on Population and Family Planning is adopted.
2004
The National Population and Family Planning Commission call for a nationwide effort to "create a favorable environment" for girls. Daughters-only or one-son families to be given bonuses and privileges.
2007
Rural parents get annual payment aged 60 if they have only one child or two girls.
ONE-CHILD FAMILY POLICIES
1980–2007
Minority nationalities have generally been allowed more flexibility in family size.
Sources: Mackerras, 1997; ; press reports

WHEN THE NEST IS OVERTURNED NO EGG STAYS UNBROKEN

Over 90 percent of people in China are Han Chinese. Just over 100 million people are from one of 55 officially recognized ethnic groups, known as minority nationalities.

Because of the strong presence of minority nationalities in Guangxi, Inner Mongolia, Ningxia, Tibet, and Xinjiang, these have been designated Autonomous Regions (ARs). This confers national minorities with some political and cultural rights, but, in practice, they enjoy little power. The Hakka are still waiting for minority nationality status.

Several ARs are located along China's borders, and are significant in terms of national security. Others are rich in natural resources, and vital to the country's economy. Not all ethnic minorities are comfortable within the territory of the People's Republic of China, however. There are Tibetans and minority nationalities in Xinjiang actively working for separation from China. Not surprisingly, all secessionist activities are banned.

▸▸ *see also page 108*

There are at least 100,000 Tibetans living in India and 20,000 in Nepal.

The first railway link between Lhasa in Tibet and Qinghai was completed in 2007, furthering not only tourism, but the influx of Han Chinese, who currently number over 85,000.

During the late 1990s, the Dalai Lama's demands for full independence were modified to "genuine self-rule", and have remained so despite riots and their suppression in 2008.

 Copyright © Myriad Editions

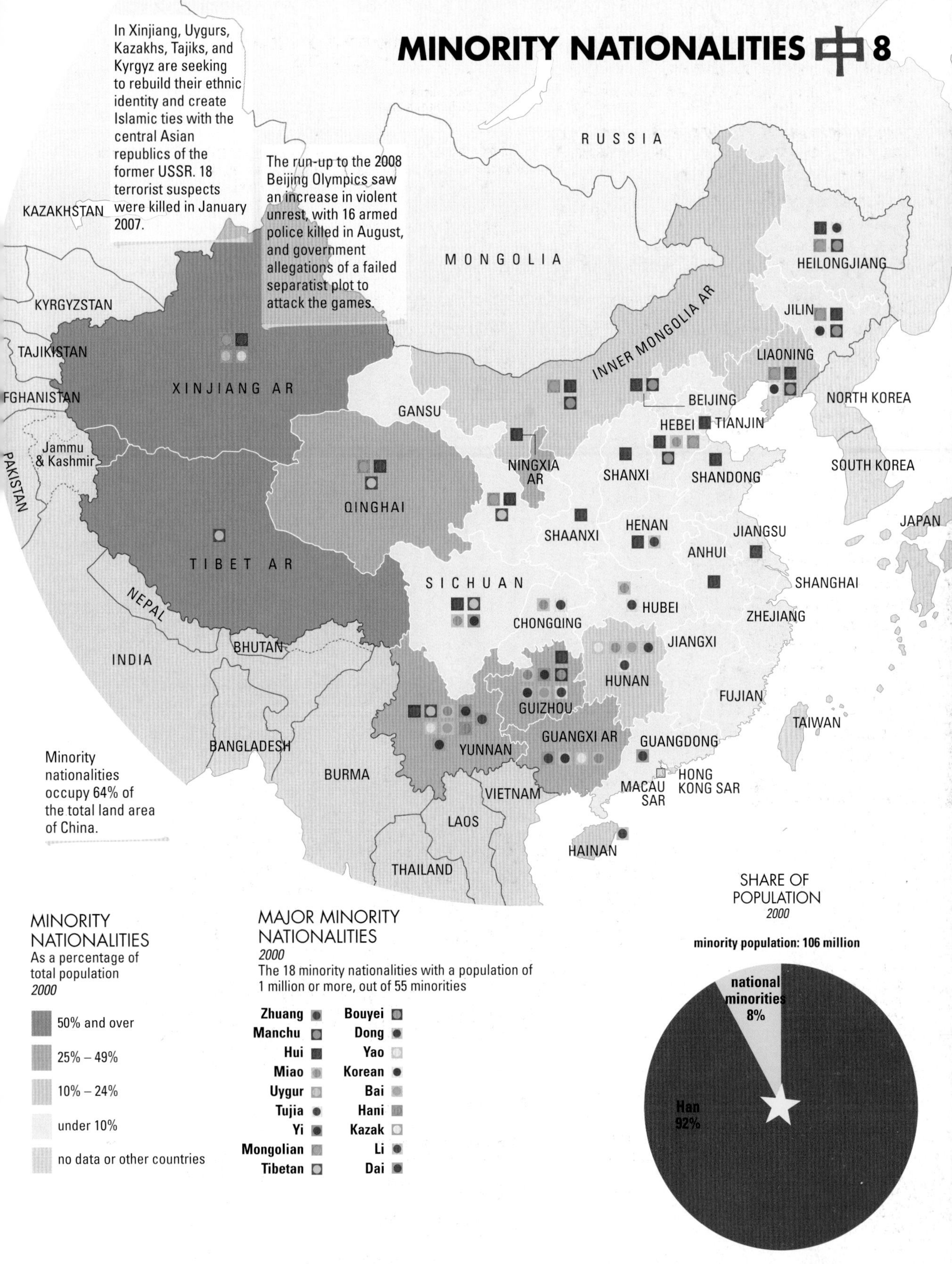
MINORITY NATIONALITIES 中 8
In Xinjiang, Uygurs, Kazakhs, Tajiks, and Kyrgyz are seeking to rebuild their ethnic identity and create Islamic ties with the central Asian republics of the former USSR. 18 terrorist suspects were killed in January 2007.
The run-up to the 2008 Beijing Olympics saw an increase in violent unrest, with 16 armed police killed in August, and government allegations of a failed separatist plot to attack the games.
Minority nationalities occupy 64% of the total land area of China.
RUSSIA
KAZAKHSTAN
MONGOLIA
KYRGYZSTAN
TAJIKISTAN
FGHANISTAN
PAKISTAN
Jammu & Kashmir
XINJIANG AR
TIBET AR
QINGHAI
GANSU
INNER MONGOLIA AR
NINGXIA AR
HEILONGJIANG
JILIN
LIAONING
BEIJING
TIANJIN
HEBEI
SHANXI
SHANDONG
NORTH KOREA
SOUTH KOREA
JAPAN
SHAANXI
HENAN
JIANGSU
ANHUI
SHANGHAI
SICHUAN
CHONGQING
HUBEI
ZHEJIANG
JIANGXI
HUNAN
GUIZHOU
FUJIAN
TAIWAN
YUNNAN
GUANGXI AR
GUANGDONG
HONG KONG SAR
MACAU SAR
HAINAN
NEPAL
BHUTAN
INDIA
BANGLADESH
BURMA
VIETNAM
LAOS
THAILAND
MINORITY NATIONALITIES
As a percentage of total population
2000
50% and over
25% – 49%
10% – 24%
under 10%
no data or other countries
MAJOR MINORITY NATIONALITIES
2000
The 18 minority nationalities with a population of 1 million or more, out of 55 minorities
Zhuang
Manchu
Hui
Miao
Uygur
Tujia
Yi
Mongolian
Tibetan
Bouyei
Dong
Yao
Korean
Bai
Hani
Kazak
Li
Dai
SHARE OF POPULATION
2000
minority population: 106 million
national minorities 8%
Han 92%

LET SOME PEOPLE GET RICH FASTER THAN OTHERS

Although China can proudly boast of lifting 250 million people out of poverty, it remains among the most unequal societies in the world.

There is a deep-seated inequality in terms of natural resources and financial wealth between the Central and Eastern regions, which are well-placed for economic development, and the mainly rural Western hinterlands. The countryside was the first to benefit from economic reforms begun in 1978, but between1985 and 1995 the gap between urban and rural widened again. Spending in rural households has averaged around a quarter of that of urban households since the mid-1990s.

The Party-State has been addressing this imbalance and is aiming to increase the income of rural populations by cutting taxes, cracking down on corrupt local officials and higher grain prices, and by improving farming methods.

▸▸ *see also page 109*

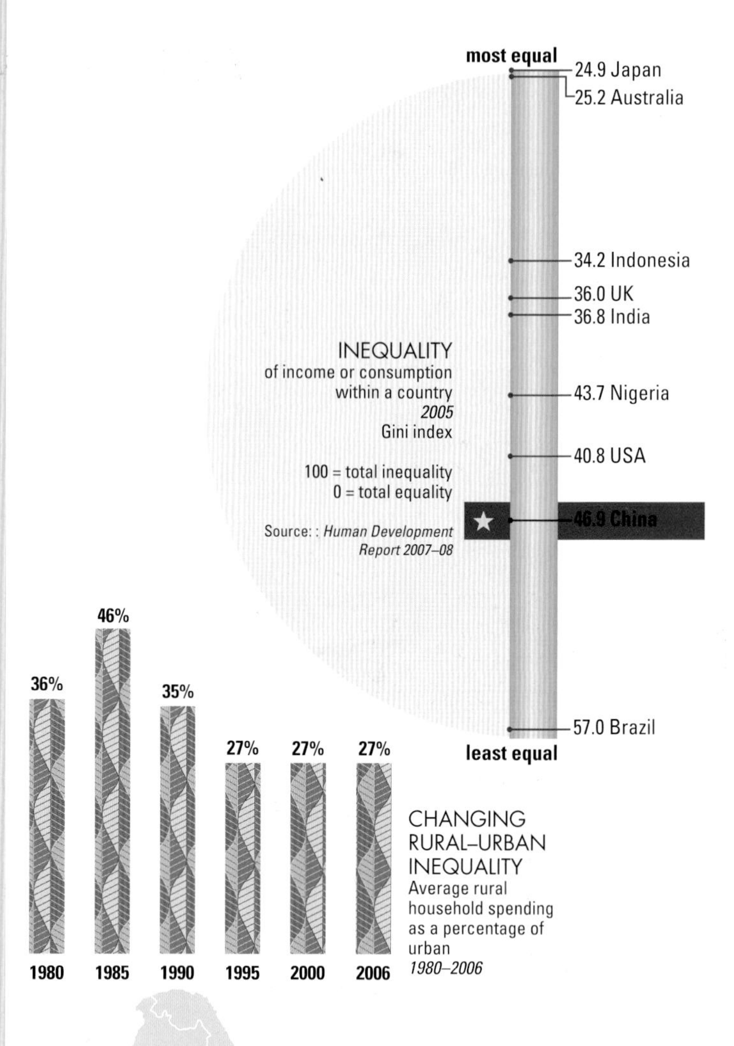

CHANGING RURAL–URBAN INEQUALITY
Average rural household spending as a percentage of urban
1980–2006

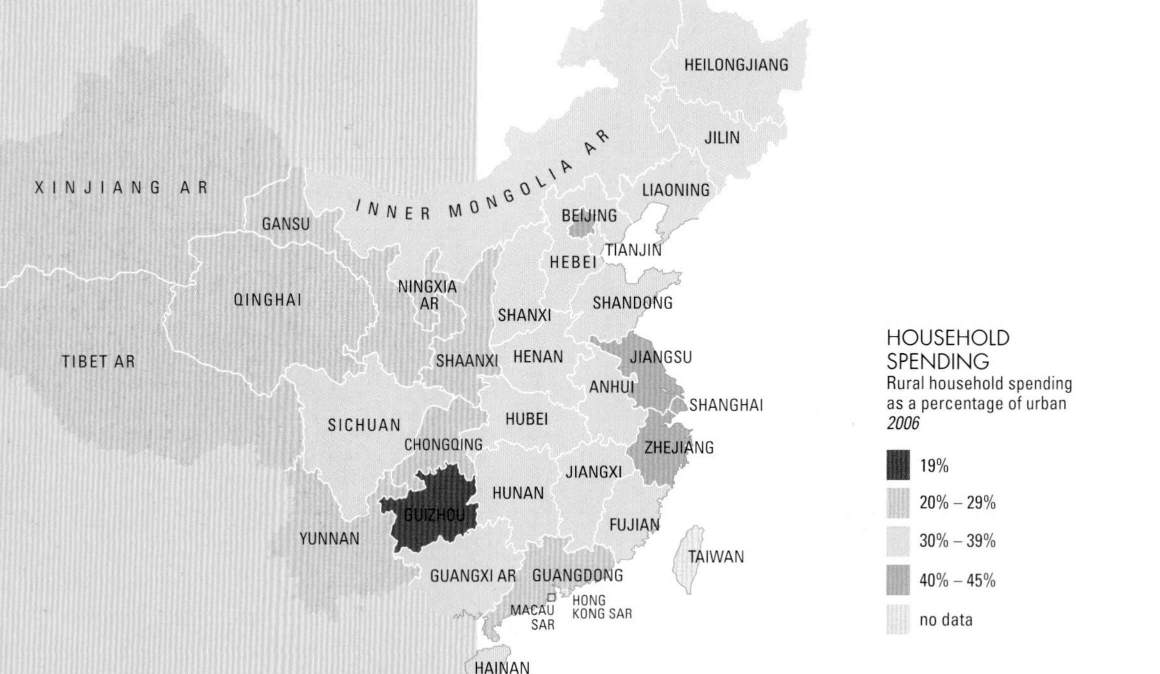

HOUSEHOLD SPENDING
Rural household spending as a percentage of urban
2006

 Copyright © Myriad Editions

Regions that are predominantly rural lag behind more urbanized regions in terms of life expectancy and provision of education.

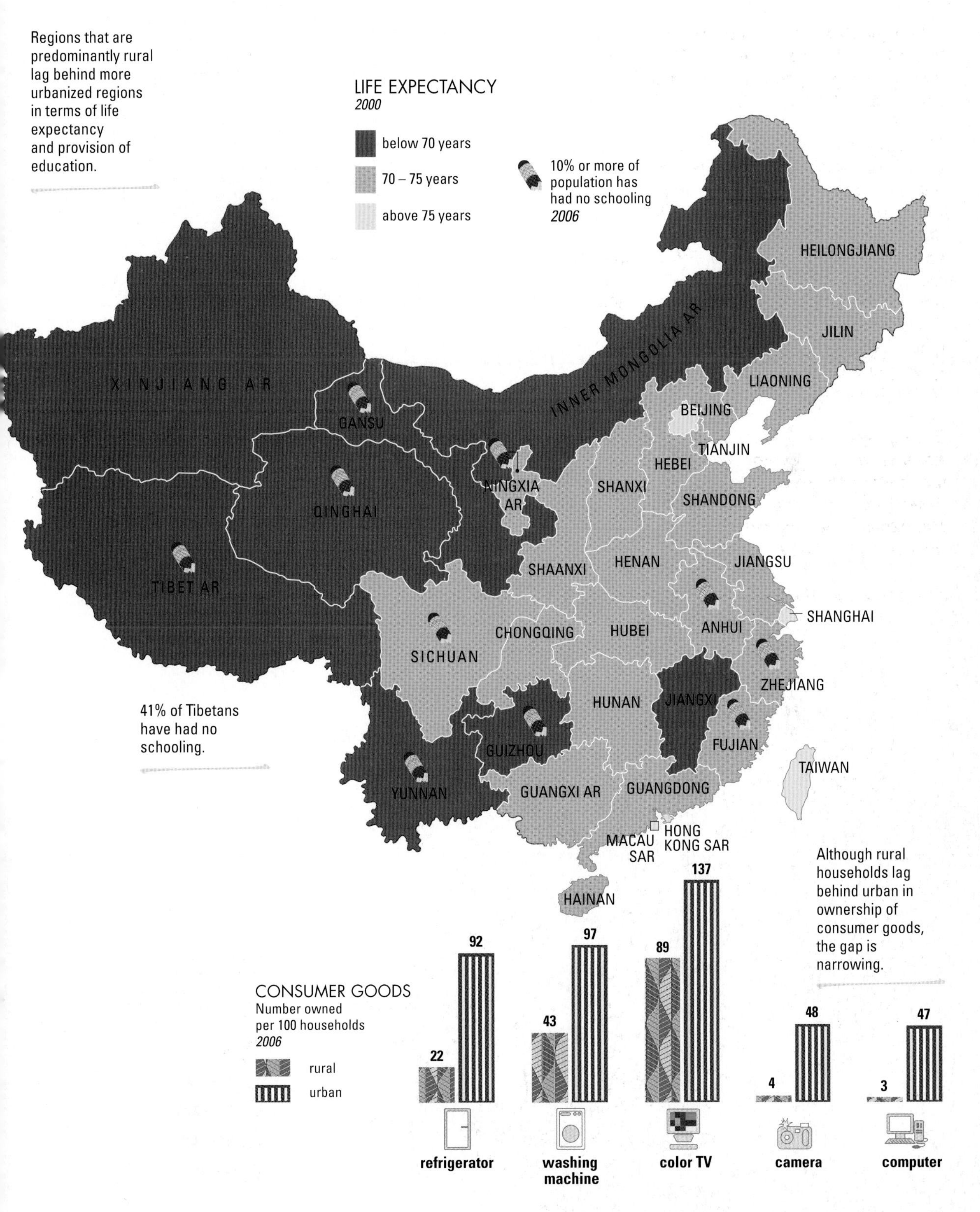

41% of Tibetans have had no schooling.

Although rural households lag behind urban in ownership of consumer goods, the gap is narrowing.

Data source: *China Statistical Yearbook 2007* unless stated otherwise

One third of the growth in the world's economy in 2008 was contributed by China.

Part Three
THE ECONOMY

THE CHINESE ECONOMY is central to global financial development, but is neither predictable nor static. Given the instability in global capitalism and the apparent fragility of the banking sector, as evidenced by the 2008 crash, the only certainty about the economy in China is that nothing is as certain as it ideally should be. Over the past 30 years, China has shifted from a command economy to the expansion of corporate capitalism and investment, and latterly to a beckoning future as the world's leading knowledge and innovation center. In early 2008, the Premier indicated that the social structure needed time to keep pace with the speed of growth, which is having a profound impact on social and environmental conditions.

Starting perhaps with the town and village enterprises of the 1980s, China has striven to become an entrepreneurial environment, with individuals increasingly mindful of the shrinking welfare state and the growth of private wealth, and of their place in that scenario. Wealth has created a demand for private cars, luxury goods and high-quality housing. For the nation as a whole it has afforded the chance to engage in large infrastructure projects, connecting the nation with high-speed communications systems, transport, and, of course, symbolic undertakings such as the Asian Games, the Olympic Games and the Shanghai World Expo.

In the wake of the Tibetan uprising in 2008, Wen Jiabao again emphasized the need to increase infrastructure links between the autonomous regions in the west and the eastern seaboard provinces. He is of a new generation of politicians who recognize that a strong state and economy are underpinned by the soft power of media and events. Thus, state investment in good intra-provincial links not only increase domestic trade, but also shores up internal stability, through the movement of goods, people and acceptable ideas.

China's formal entry to the world economy was marked by its accession to the World Trade Organization in 2001. At the time the Party-State saw the WTO as a "wrecking ball" that would sort out the weak from the strong with as little political unrest as possible. The benefits of an international system include the opportunity for the State to blame overseas issues for failure and local wisdom for success. There have been several landmark moments since 2001, including the meltdown of the markets late in 2008, which had begun to shake stock markets in Shanghai several months earlier. Although the Chinese stock market does not attract personal savings to the extent it does in the West, and although the Shanghai market has a tendency to volatility, the failure of the markets in 2008 was seen as yet another example of anti-Chinese behavior by the West in the Olympic year.

Outbursts of nationalism are symptomatic of the challenges the government faces in managing the economy. It wishes to retain total political power (largely based on nationalism over other ideologies) but without letting the country slip either into isolationist panic or internal scuffles over financial and social rights. The second, and related, challenge is to balance the needs of the population with a sustainable level of growth, especially in years of recession amongst China's major trading partners. It is estimated that China requires at least 7 percent annual growth simply to keep the domestic population on an even financial keel, and more than that to sustain important reforms in education access, rural development and agricultural expansion.

HORSES NEVER EAT THE GRASS BEHIND THEM

Between 1990 and 2007, China's economy has grown by an average of 10 percent a year.

In order to create a more balanced economy across China's provinces, the government has been encouraging enterprises to move inland and to "open up to the West", with some success. But although labor is cheaper in these areas, there are fewer skilled laborers, and transportation costs are higher. Foreign investors are so far showing little inclination to venture beyond the coastal provinces.

The state's massive investment in infrastructure has had a major impact on China's exceptional rate of economic growth, which has, in turn, brought a significant increase in central government revenue through taxation. From 2006 to 2007, tax revenue rose by more than 30 percent. However, inflation also increased dramatically, mainly fueled by increases in the price of food.

see also page 110

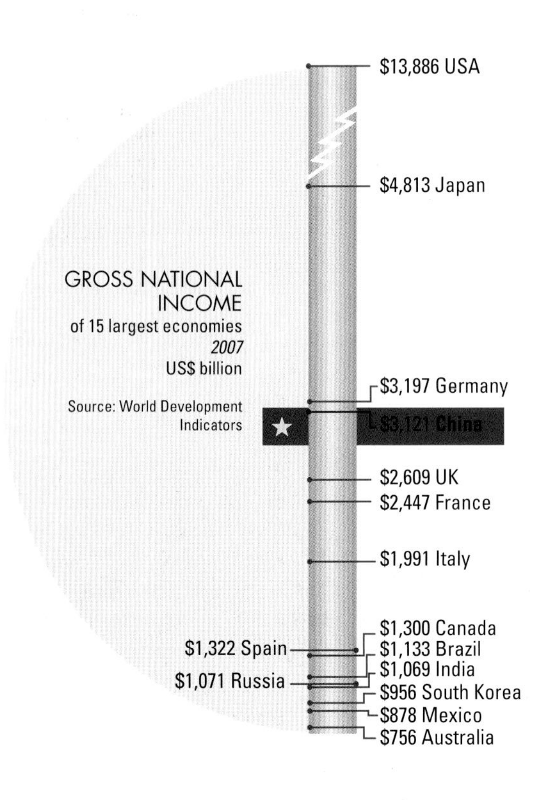

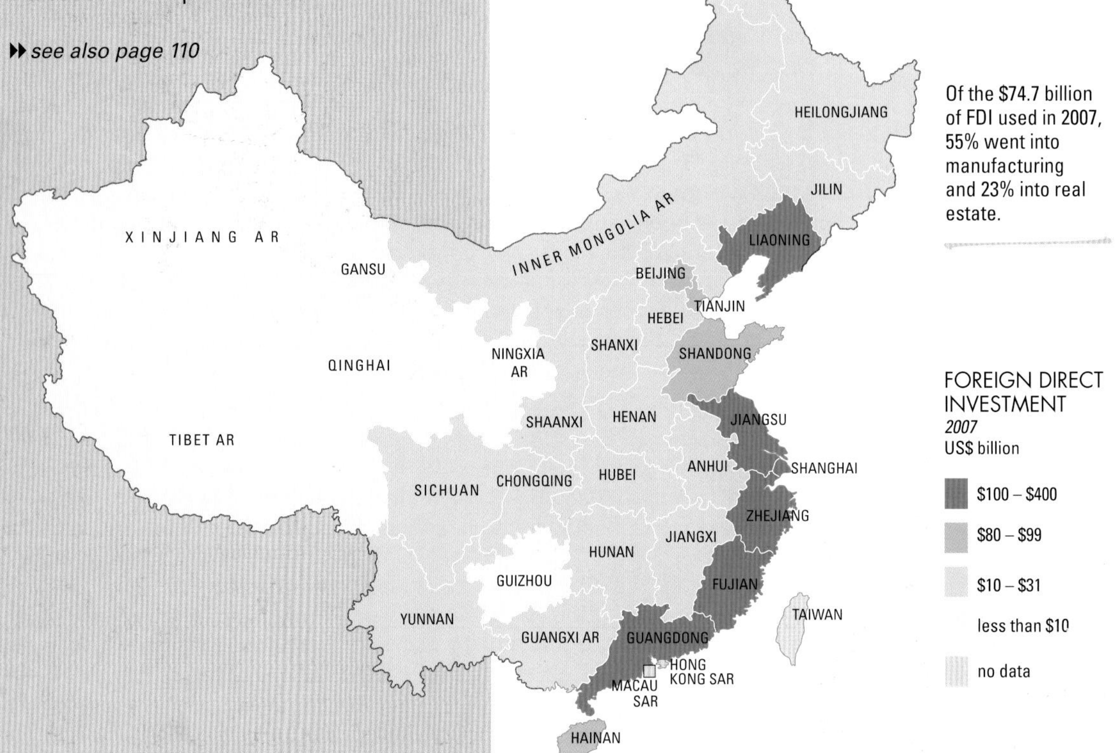

Of the $74.7 billion of FDI used in 2007, 55% went into manufacturing and 23% into real estate.

FOREIGN DIRECT INVESTMENT
2007
US$ billion

- $100 – $400
- $80 – $99
- $10 – $31
- less than $10
- no data

 Copyright © Myriad Editions

In response to the global economic downturn in 2008, China's government announced a spending package worth 4 trillion yuan to boost domestic demand.

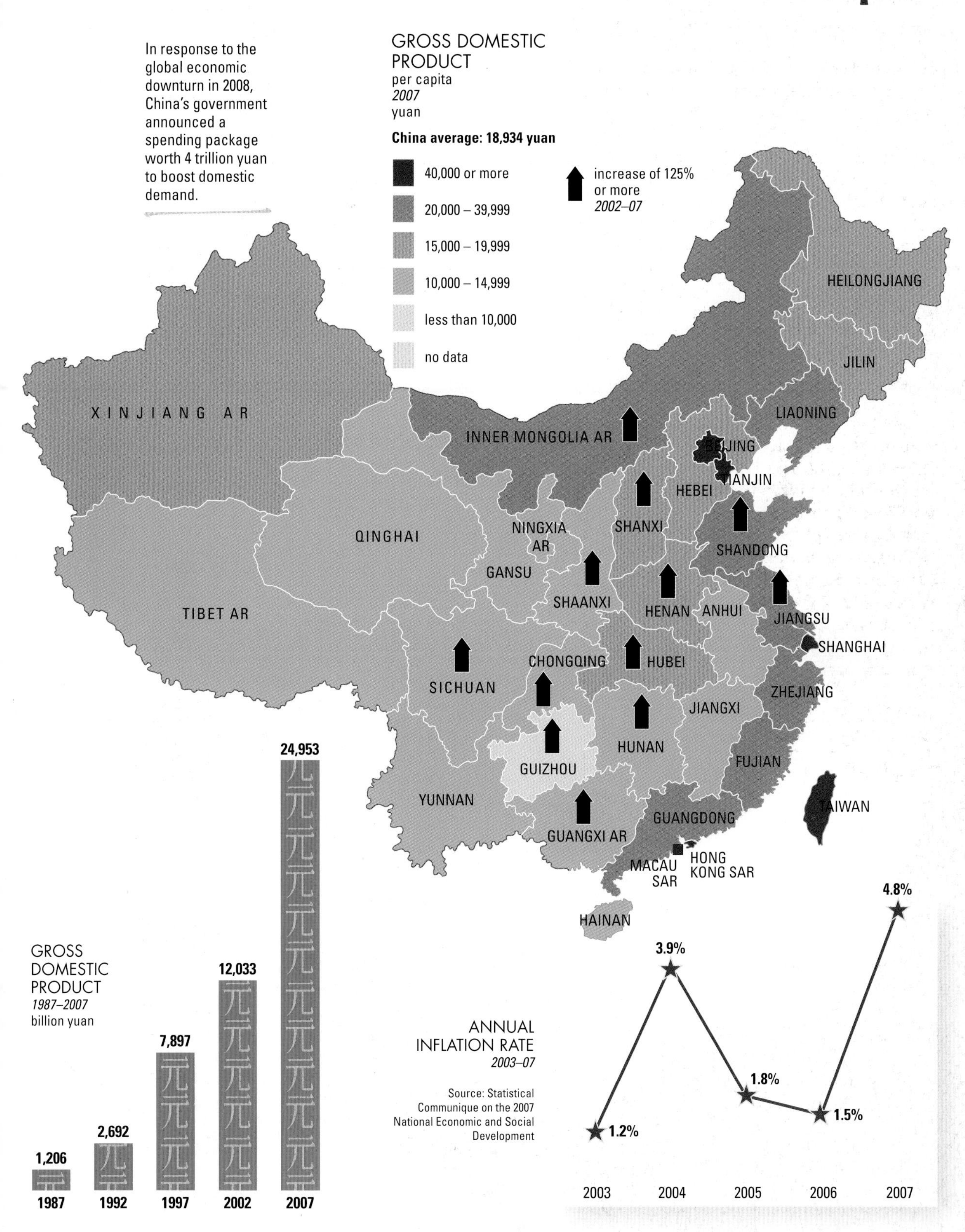

BETTER THE HEAD OF A CHICKEN THAN THE TAIL OF AN OX

The entrepreneur, scion of the market-oriented economy, is fêted in popular self-help manuals and potted biographies in China's bookshops.

The re-organization and closures of state owned enterprises, the opening of China's economy to foreign investment, and the large-scale privatization of housing has created opportunities for an "entrepreneurial class". This group comprises both newly enriched entrepreneurs, and those who have used their business acumen and political connections to advantage in the reformed economy.

Following Deng Xiaoping's dictum "Get Rich First", peasants moved out of grain production into specialist and luxury food production, or set up town and village enterprises. The area of land used for vegetable growing increased by 80 percent. Many of those who prospered became village leaders, chairing village enterprises management committees.

The dictum "two classes, one stratum" no longer holds, as workers and peasant-farmers alike, but especially domestic migrant workers, are relegated to the bottom of the pile. Class divisions in China are stark. The urban rich are accumulating substantial wealth, while aspirational urban workers struggle to maintain access to costly services and housing.

Meanwhile, the poor are caught in a cycle of deprivation. Their plight is exacerbated by the value of land, particularly for development and industrial farming. Increasing legal flexibility allows large enterprises to appropriate small plots, rendering their previous incumbents tenants, or landless. The State's concept of "harmonious society" is designed to address the growing disparities amongst its people.

▸▸ *see also page 110*

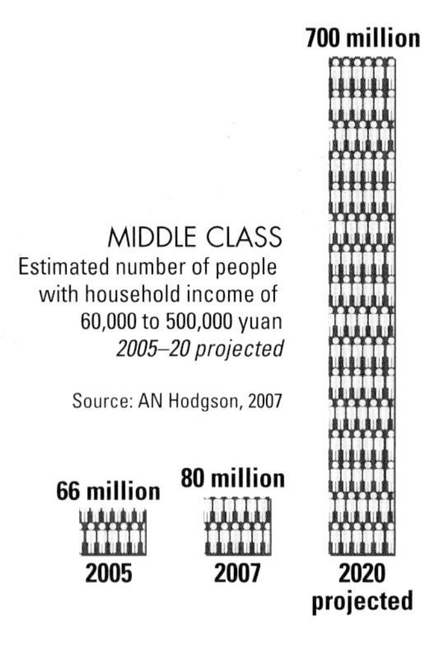

SAVINGS DEPOSITS
Amount saved
1990–2006
billion yuan

1990	1995	2000	2003	2007
712	2,966	6,433	10,362	17,253

20% of Chinese entrepreneurs are women. In 2006, a woman topped China's rich list for the first time.

China's top 100 philanthropists donated $1.8 billion to charity between 2003 and 2008.

CLASS STRUCTURE
By status as adapted by the Chinese Academy of Social Sciences
2004

1	State and society leaders
2	CEOs
3	Entrepreneurs
4	Professionals
5	Clerks
6	Shopkeepers
7	Commercial services staff
8	Industrial workers
9	Agricultural workers
10	Non-employed, unemployed, and semi-employed

 Copyright © Myriad Editions

In 2006, 90% of real-estate development was privately funded from domestic sources, compared with only 37% in 1988.

In 2006, only 5% of real-estate investment was in affordable housing. In 1999, it was 17%.

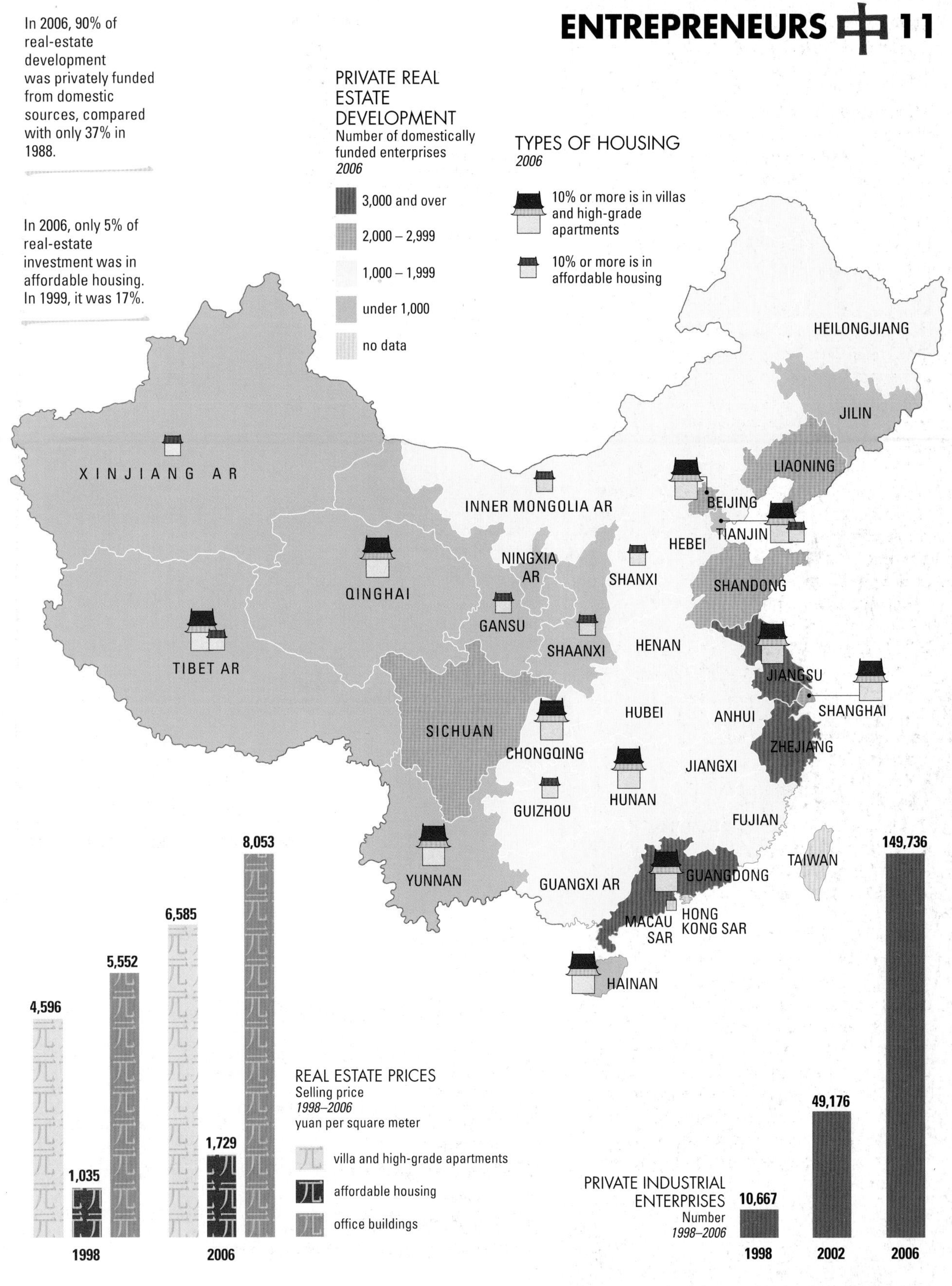

Data source: *China Statistical Yearbook 2007* and *2008* unless stated otherwise

SEEING NO OX AS A WHOLE

There are at least 8 million new entrants a year into China's labor market.

While creating jobs is, and will remain, a daunting task, it is not just the number of entrants that is a concern, but also how to ensure that they possess the appropriate skills needed for China's continued development.

A popular saying is that there are three genders in China, male, female and those with foreign doctorates. There are two even more potent dividers: those with the potential and training to work in an advanced, mixed economy, and those who can only work in unskilled and low-paid jobs with little hope for advancement or long-term security. Wage levels across the nation are highest in the metropolitan centers on the eastern seaboard.

There are anomalies in poorer regions such as Tibet, however, where inward migration of skilled labor and entrepreneurs is actively encouraged by State subsidies.

▸▸ see also page 111

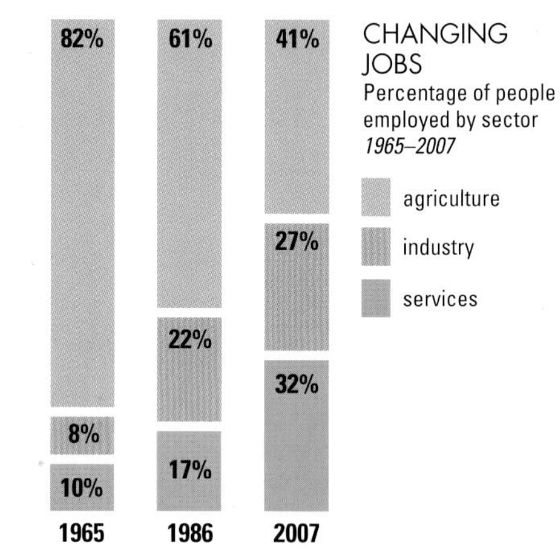

CHANGING JOBS
Percentage of people employed by sector
1965–2007

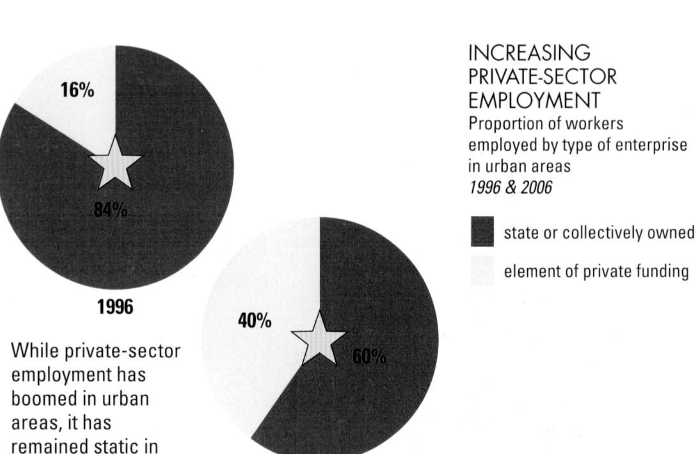

INCREASING PRIVATE-SECTOR EMPLOYMENT
Proportion of workers employed by type of enterprise in urban areas
1996 & 2006

While private-sector employment has boomed in urban areas, it has remained static in rural areas, where 90% of people are employed in township and village enterprises.

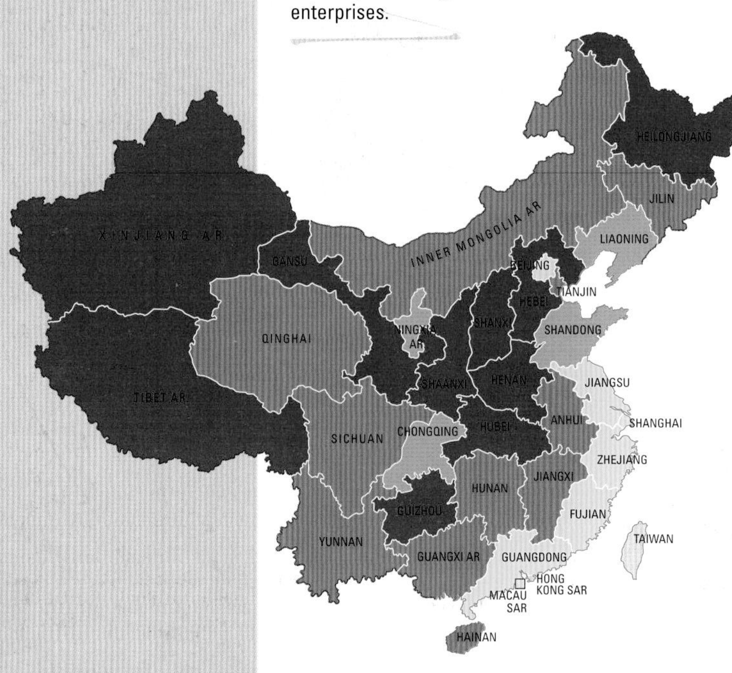

PRIVATE EMPLOYMENT
Percentage of people employed in urban enterprises with element of private funding
2006

 Copyright © Myriad Editions

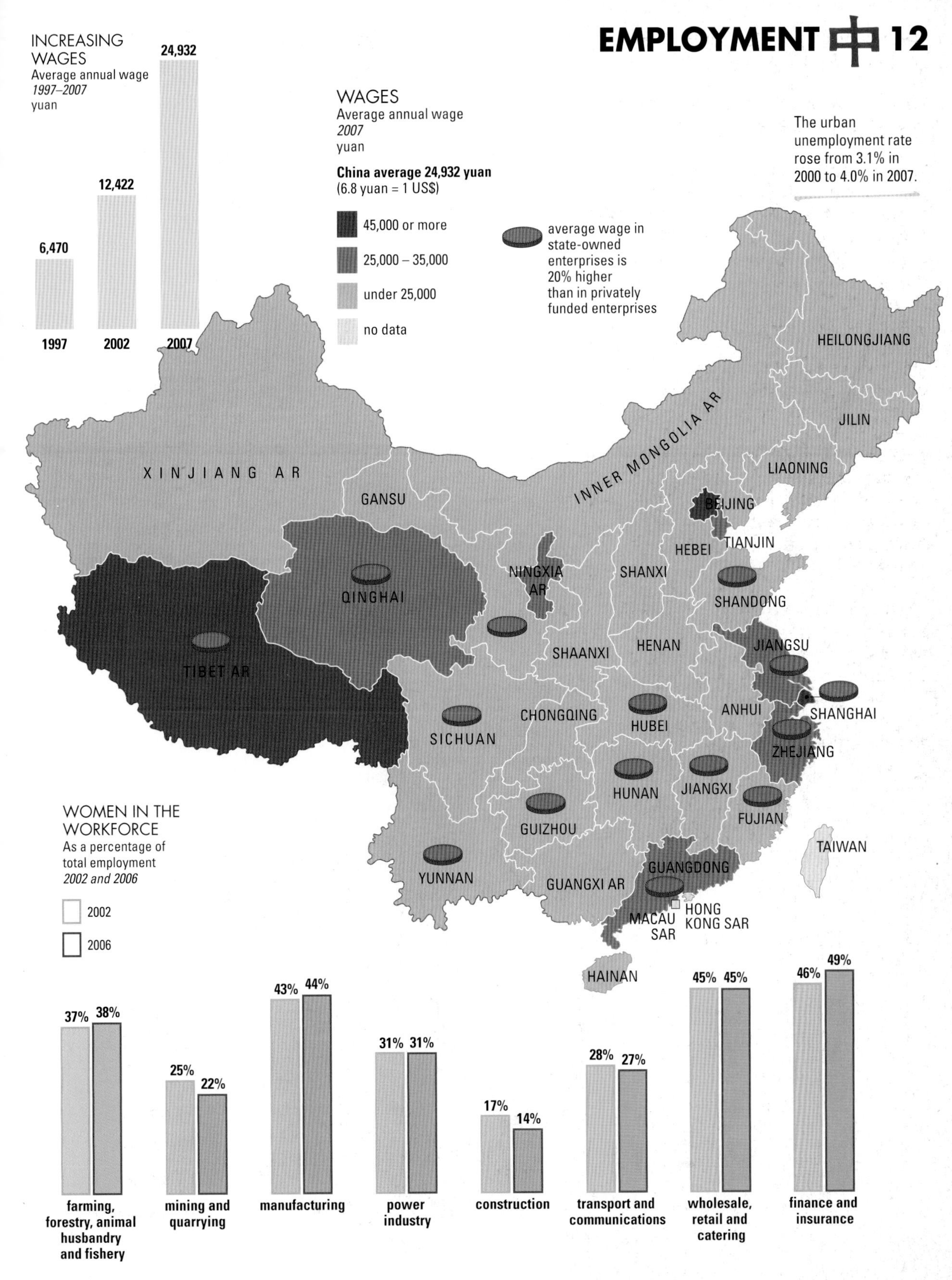

Data source: *China Statistical Yearbook 2007* and *2008* unless stated otherwise

THE THUNDER IS HUGE BUT THE RAINDROPS ARE MANY

Agriculture provides a declining share of China's GDP.

While this decline is in line with trends in other industrializing countries, ensuring a continuing and adequate food supply for the population presents the Chinese government with a particularly daunting logistical challenge.

Government investment has targeted agricultural modernization in an attempt to revive rural confidence and productivity. Animal husbandry (mainly factory farming) and production of corn for animal feed and crops vital to the food processing industries are booming, in large part due to the increased emphasis on meat and dairy produce in the Chinese diet. The amount of grain produced for human consumption was in decline, but is now climbing back towards its mid-1990s level.

In 2007, as a result of the harmonization policy, a 2,600-year-old agricultural tax was abolished, along with a host of other taxes imposed on the rural population. The following year, the government discussed a new land-tenure policy designed to encourage smallholders to create larger and more productive farms through land-lease arrangements. Such a move would affect up to 800 million people still classified as farmer-peasants.

▸▸ *see also page 112*

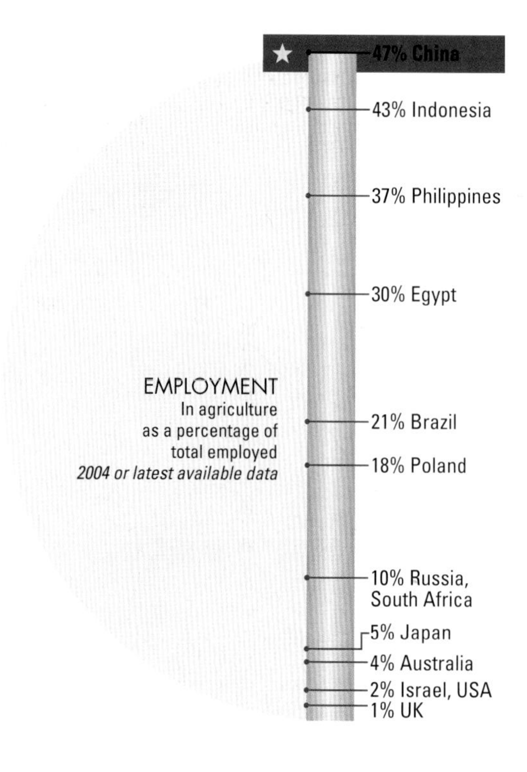

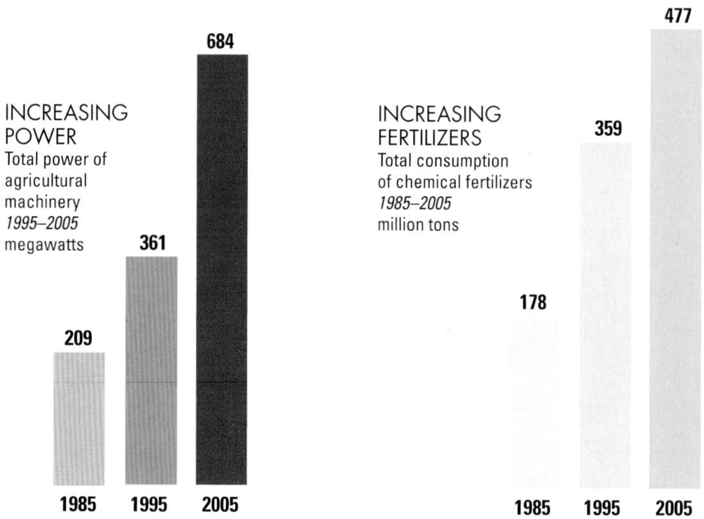

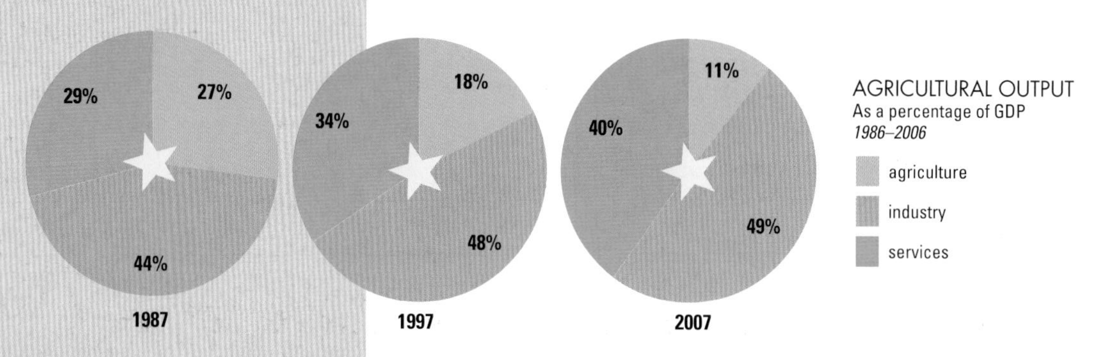

 Copyright © Myriad Editions

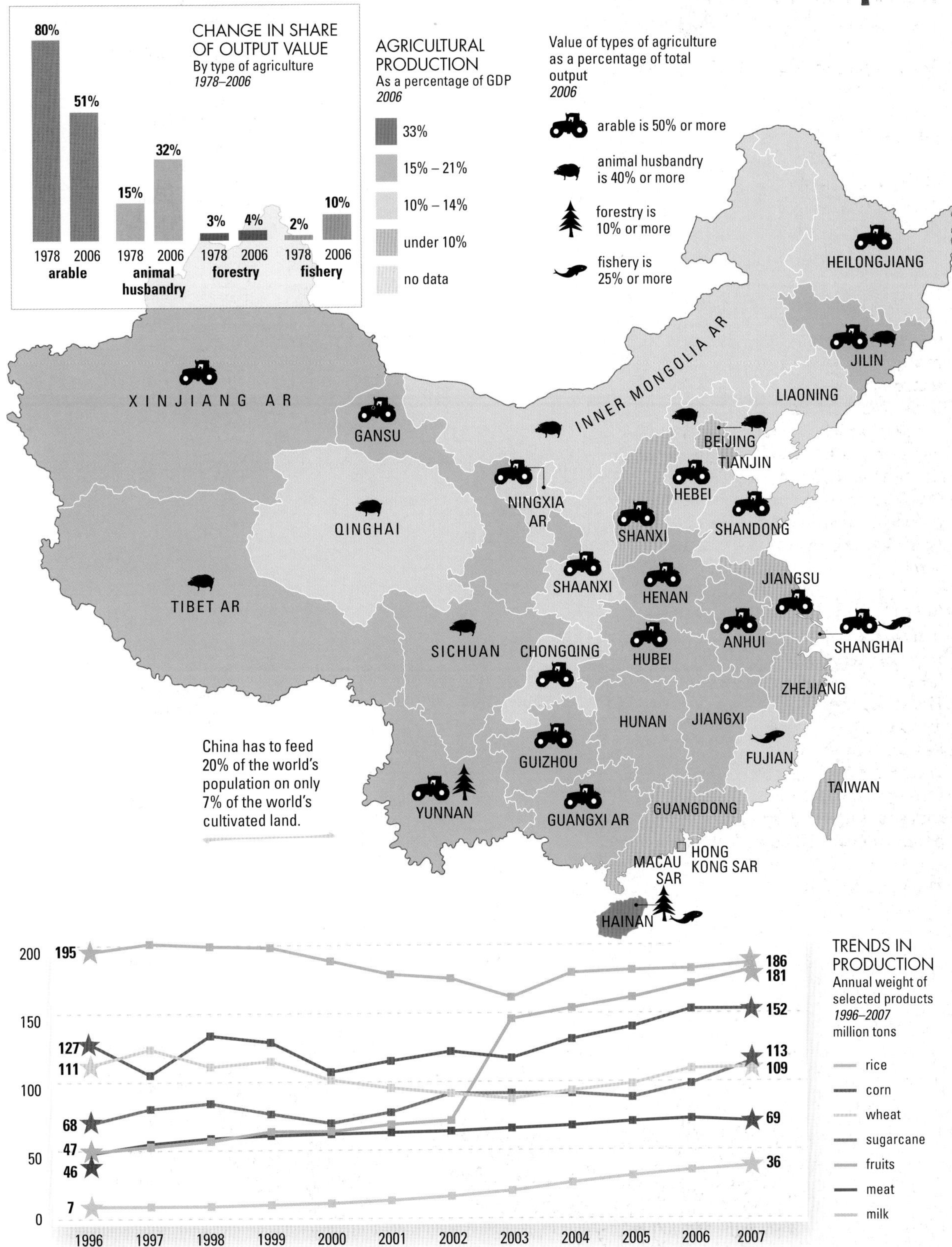

Data source: *China Statistical Yearbook 2007* and *2008* unless stated otherwise

LIKE A SILKWORM EATING THROUGH A LEAF

China's industrial output is expanding, with goods produced both for export and in the hope of building domestic demand.

The highest growth in production is clustered around three key consumer goods: PCs, cell phones and cars, all three of which have a detrimental impact on the environment.

The next fastest-growing commodity is cigarettes, the negative impact of which will be felt on the nation's health and health services in the decades to come.

China's car production in 2007 was the third greatest in the world, in part to meet the demand for the passenger vehicles that are clogging China's roads. Cell-phone manufacture includes production of global brands but also of home brands designed for local distribution. PC manufacturers have in some cases managed to globalize their brand names, with companies such as Lenovo now recognizable to users all over the world.

▸▸ *see also page 112*

EMPLOYMENT
In industry as a percentage of total employed
2004 or latest available data

30% Russia
29% Poland
28% Japan
25% South Africa
23% China
22% Israel, UK
21% Australia, Brazil, USA
20% Egypt
18% Indonesia
15% Philippines

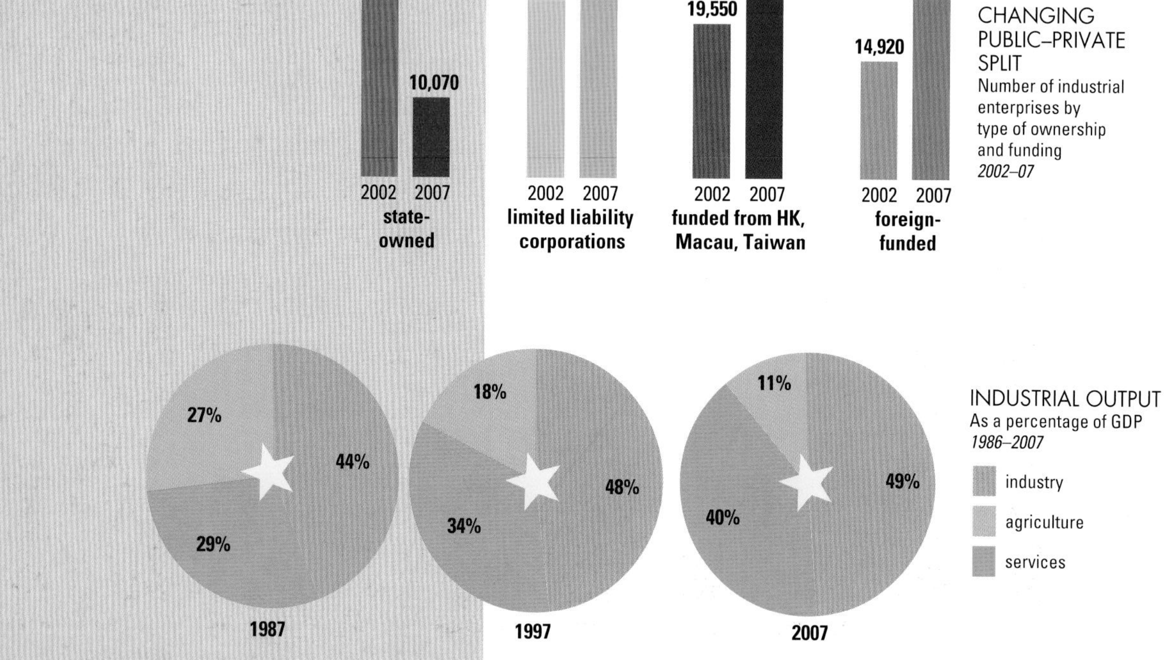

CHANGING PUBLIC–PRIVATE SPLIT
Number of industrial enterprises by type of ownership and funding
2002–07

INDUSTRIAL OUTPUT
As a percentage of GDP
1986–2007

 Copyright © Myriad Editions

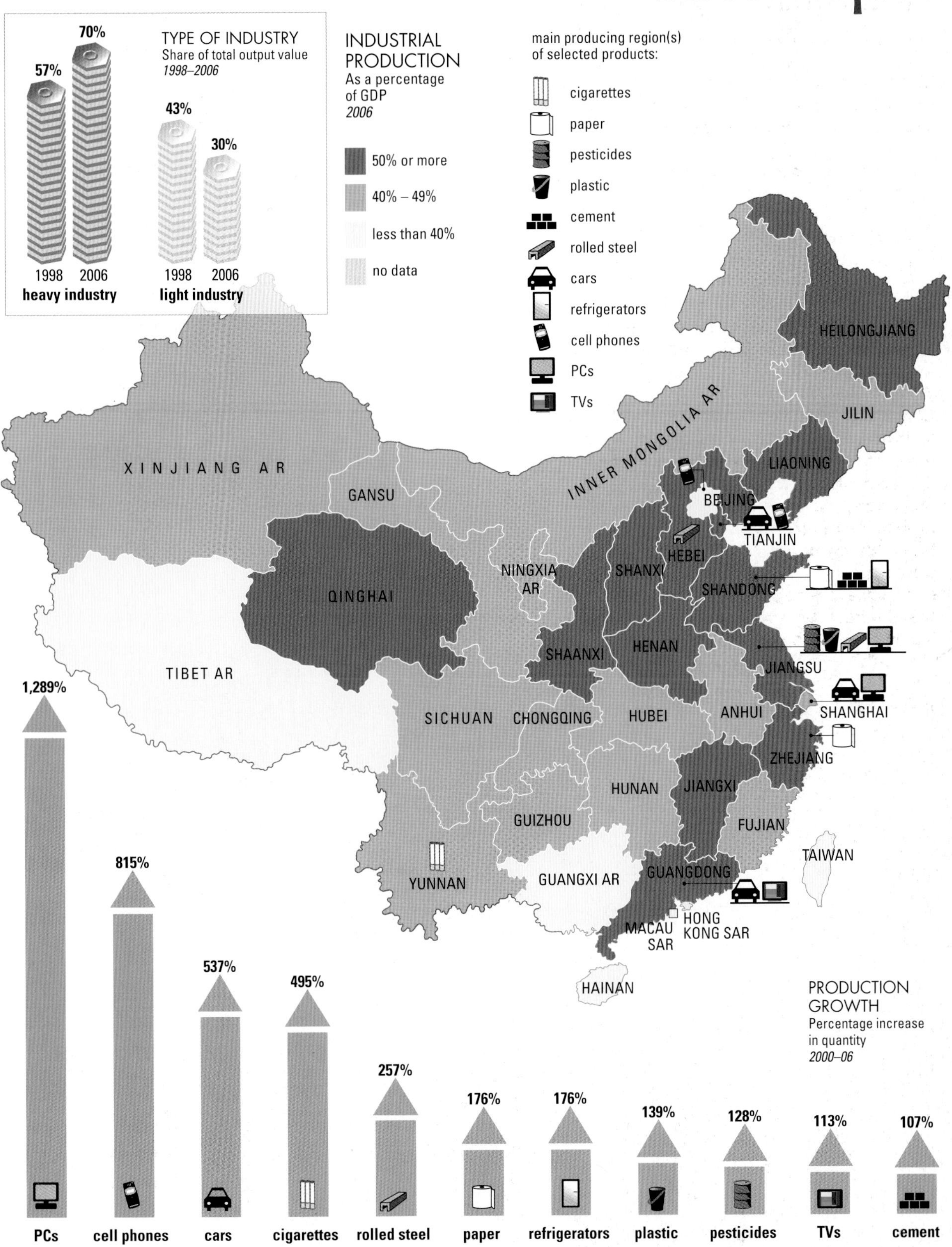

Data source: *China Statistical Yearbook 2007* and *2008* unless stated otherwise

IT DOESN'T MATTER WHAT COLOR THE CAT IS, SO LONG AS IT CATCHES THE MICE

Over the last 20 years, the share of GDP contributed by the services sector has grown by 10 percent.

Most of this growth would seem to be at the expense of agriculture, but the gap between industry and services is also shrinking, and China's economy is looking increasingly like that of the industrialized nations. The challenge to the government and to China's business community is to maintain a healthy level of production, which fuels growth, whilst also developing a strong and well-regulated infrastructure in the service areas of finance, education and training, welfare, real estate, creative industries, and leisure and hospitality.

The chronic instability of global markets may slow the growth of services in the immediate future. Whilst the escalation of real-estate prices and the instability of overheated investment markets might well be curbed, the long-term needs and interests of the population lie in a sophisticated and well-managed services environment.

▸▸ *see also page 112*

EMPLOYMENT
In services as a percentage of total employed
2004 or latest available data

78% USA
76% UK
75% Australia, Israel
66% Japan
65% South Africa
60% Russia
58% Brazil
53% Poland
50% Egypt
48% Philippines
39% Indonesia
31% China

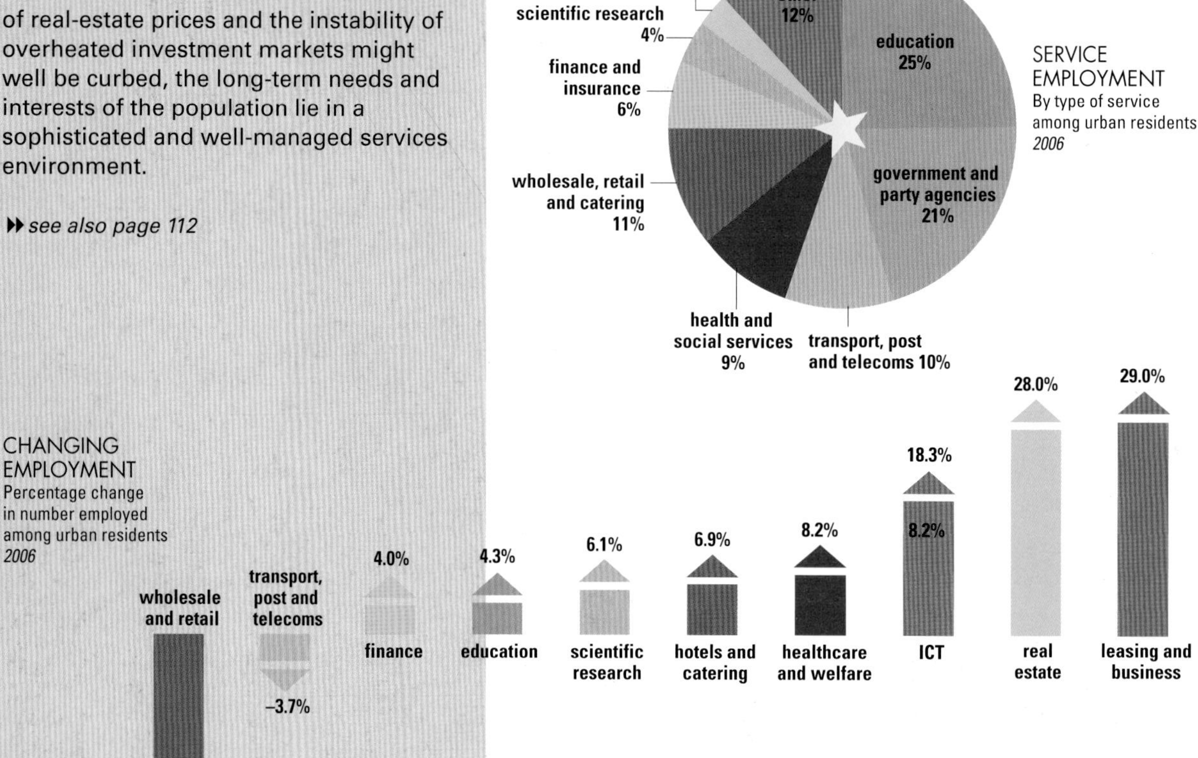

 Copyright © Myriad Editions

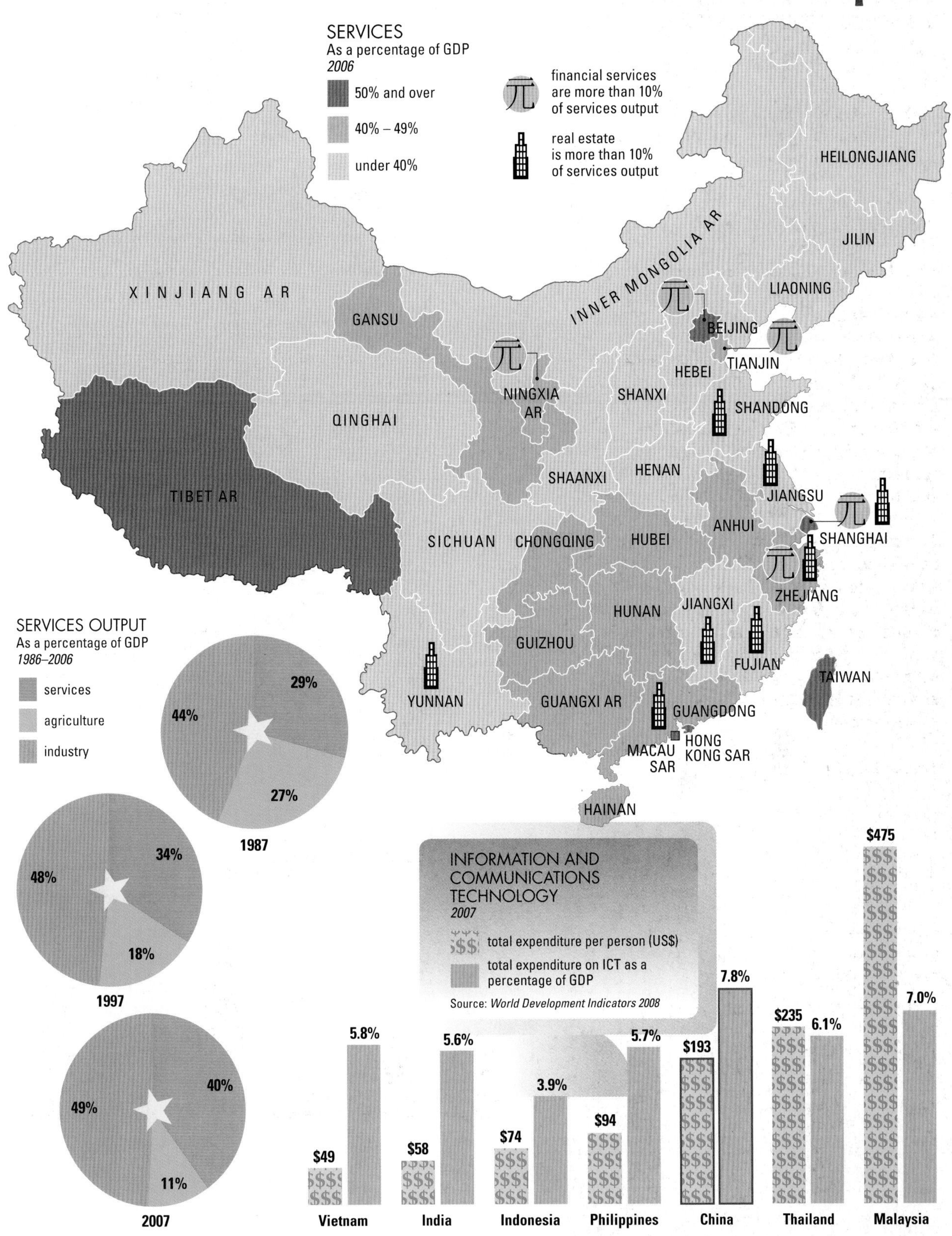

Data source: *China Statistical Yearbook 2007* and *2008* unless stated otherwise

THE GRASS STIRS AS THE WIND BLOWS

The major focus of China's tourism industry is on domestic mainland tourists and, to a lesser extent, on visitors from Hong Kong.

The money they bring to the major tourist sites to a large extent fuels the industry. Although international tourism, which includes visiting overseas Chinese, is also important, much of it is confined to major cities such as Beijing and Shanghai. And it is still secondary to the larger project of managing China's brand in order to increase the country's national status and authority in the region and across the world.

China's concern to communicate its own image to a global audience, rather than to individual and sometimes difficult tourists, is largely achieved through televized versions of major events, such as the Shanghai International Expo in 2010.The success of the 2008 Olympics was not measured in terms of the number of visitors to Beijing, which was actually lower than in August 2007 as a result of fierce visa controls, but in how effectively they conveyed to the world's audiences a convincing and overwhelming message about China's power, competence and cultural worth.

▸▸ *see also page 113*

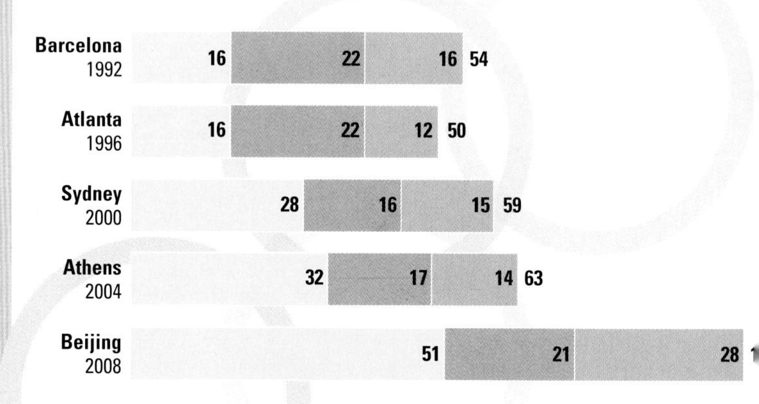

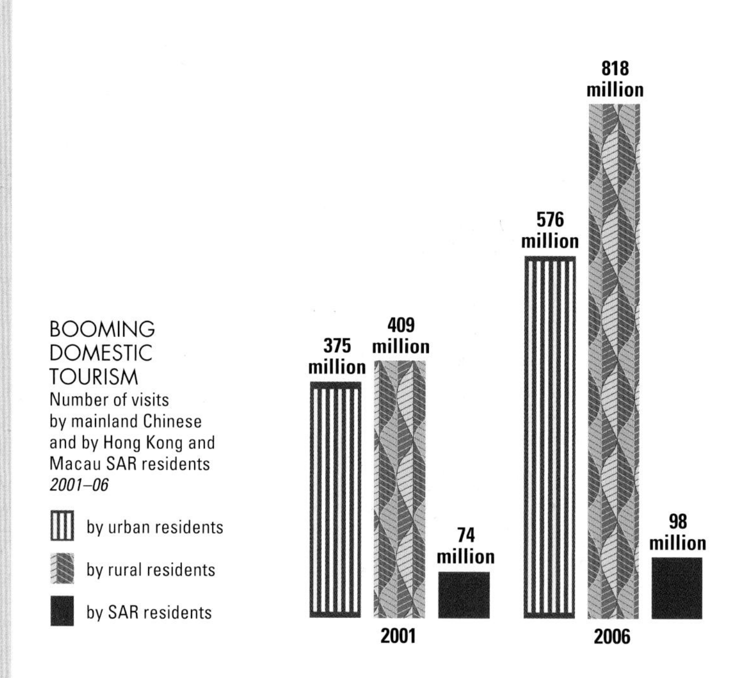

 Copyright © Myriad Editions

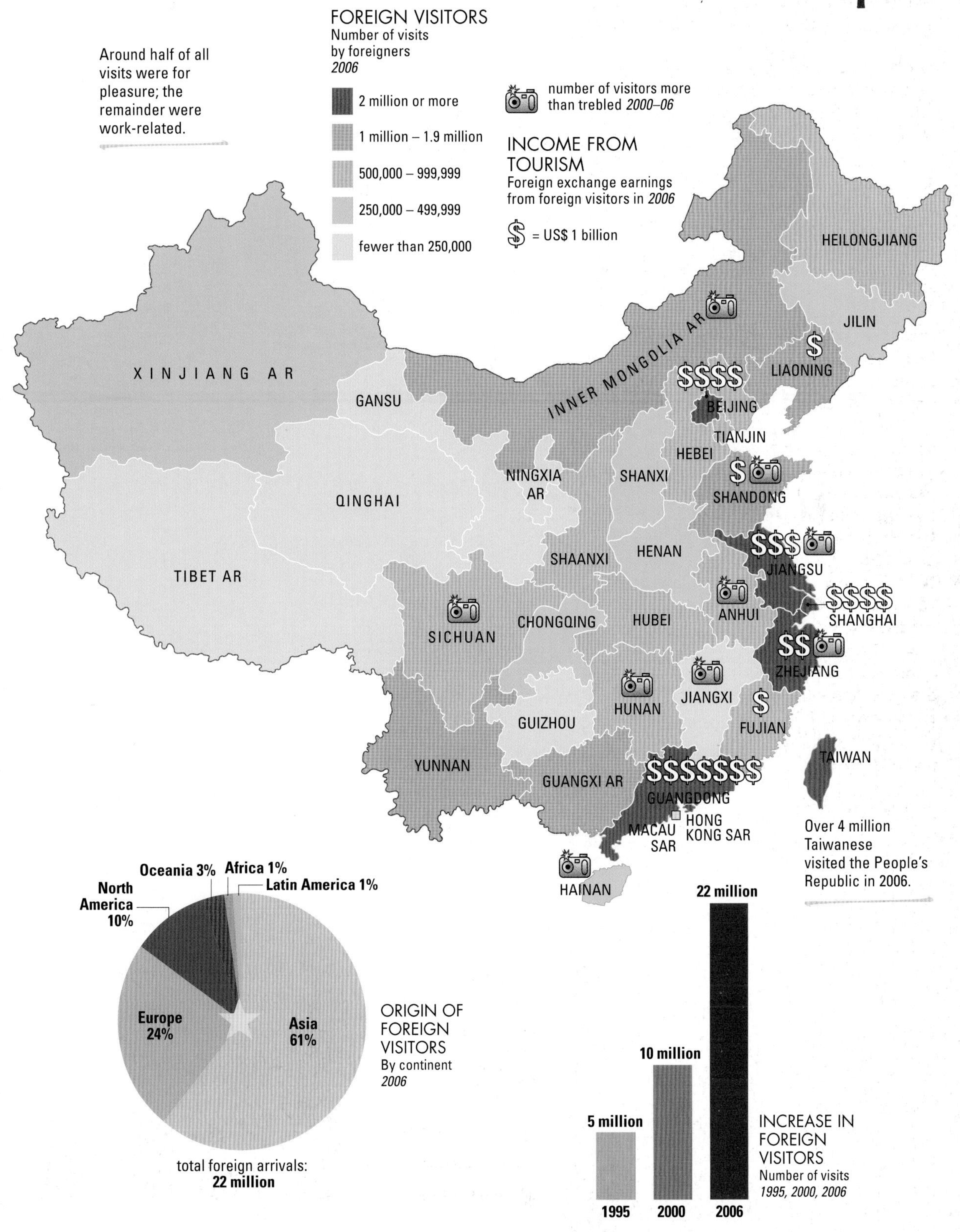

Data source: *China Statistical Yearbook 2007* unless stated otherwise

A HEAVY WEIGHT HANGS BY A HAIR

Following a period early in the decade when China's energy production was unable to keep up with demand, its capacity has soared.

China's reliance on coal has declined slightly in recent years, but it still provides 70 percent of energy. However, despite being the world's leading coal producer, China became a net importer in April 2007 – getting its supplies mainly from Australia and South America.

China ranked fifth in the world in the production of crude oil in 2007, but its demand for energy is so great that it needs to import more than 1 billion barrels a year, which it does from an increasingly wide range of countries.

China has the largest hydroelectric generating capacity in the world. Its most famous large dam, the Three Gorges, is just one among over 25,000, and the government is aiming to double hydropower generating capacity by 2020, building many more dams, including a series on the Nu (Salween) and the Lancang (Upper Mekong) rivers. This will have massive human and environmental impacts, both within China, and on its downstream neighbors in South-East Asia.

▸▸ *see also page 113*

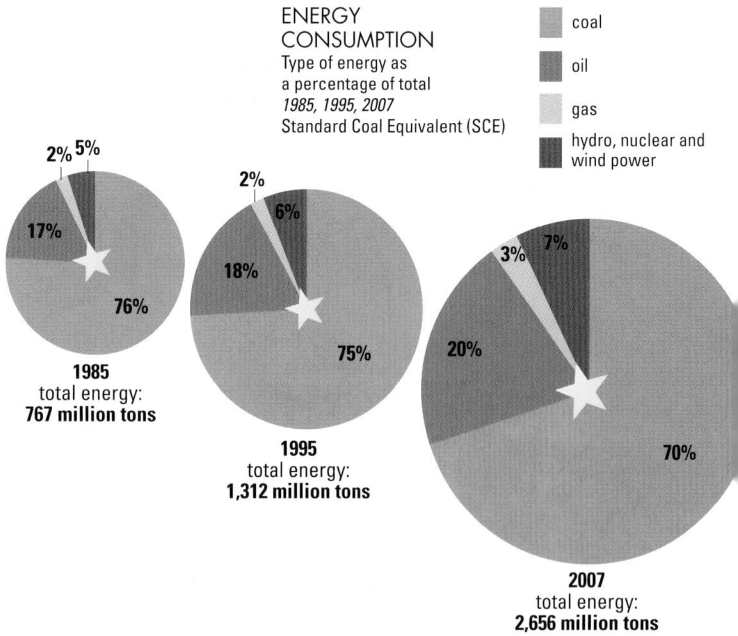

ENERGY CONSUMPTION
Type of energy as a percentage of total
1985, 1995, 2007
Standard Coal Equivalent (SCE)

coal
oil
gas
hydro, nuclear and wind power

76%
17%
2%
5%
1985
total energy:
767 million tons

75%
18%
2%
6%
1995
total energy:
1,312 million tons

70%
20%
3%
7%
2007
total energy:
2,656 million tons

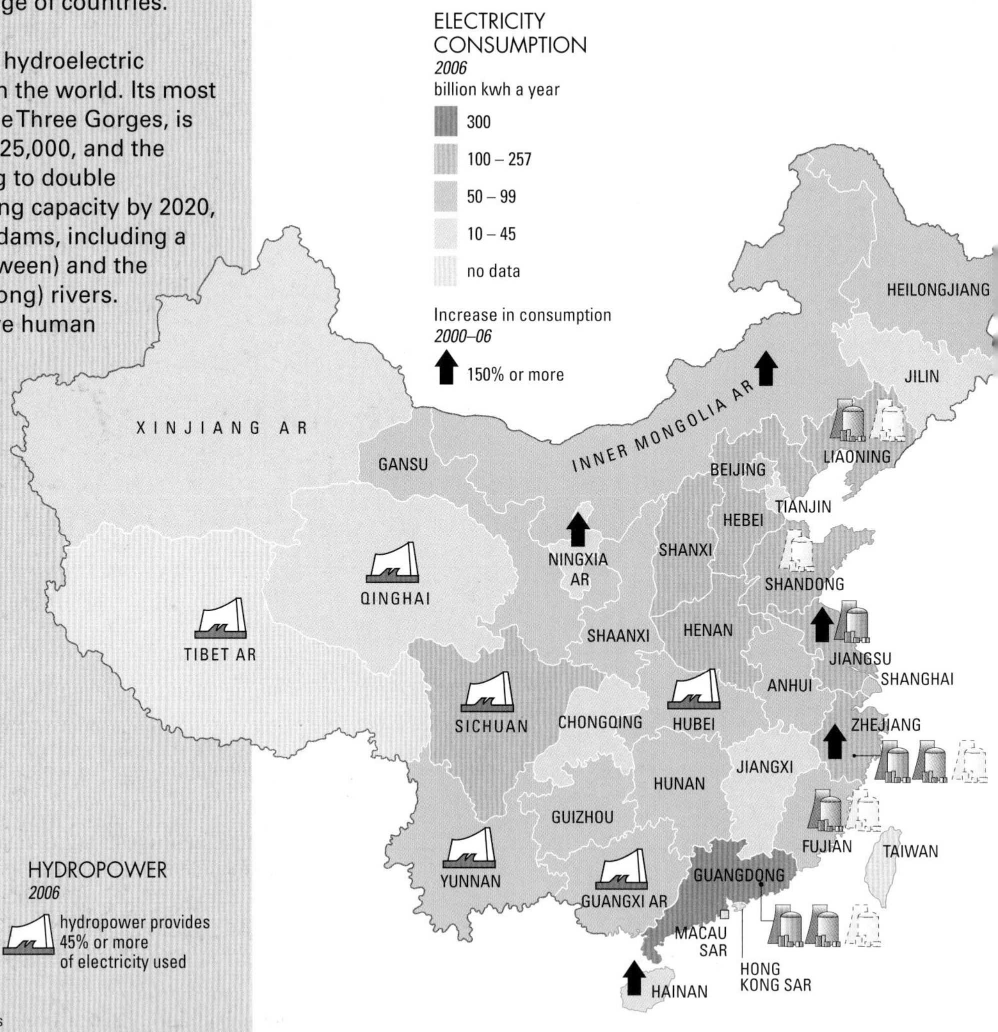

ELECTRICITY CONSUMPTION
2006
billion kwh a year
300
100 – 257
50 – 99
10 – 45
no data

Increase in consumption
2000–06
150% or more

XINJIANG AR
GANSU
INNER MONGOLIA AR
HEILONGJIANG
JILIN
LIAONING
BEIJING
TIANJIN
HEBEI
SHANXI
NINGXIA AR
QINGHAI
TIBET AR
SHANDONG
SHAANXI
HENAN
JIANGSU
SHANGHAI
ANHUI
SICHUAN
CHONGQING
HUBEI
ZHEJIANG
JIANGXI
HUNAN
GUIZHOU
FUJIAN
TAIWAN
YUNNAN
GUANGXI AR
GUANGDONG
MACAU SAR
HONG KONG SAR
HAINAN

NUCLEAR POWER
2008
operating power stations
under construction
proposed

HYDROPOWER
2006
hydropower provides 45% or more of electricity used

 Copyright © Myriad Editions

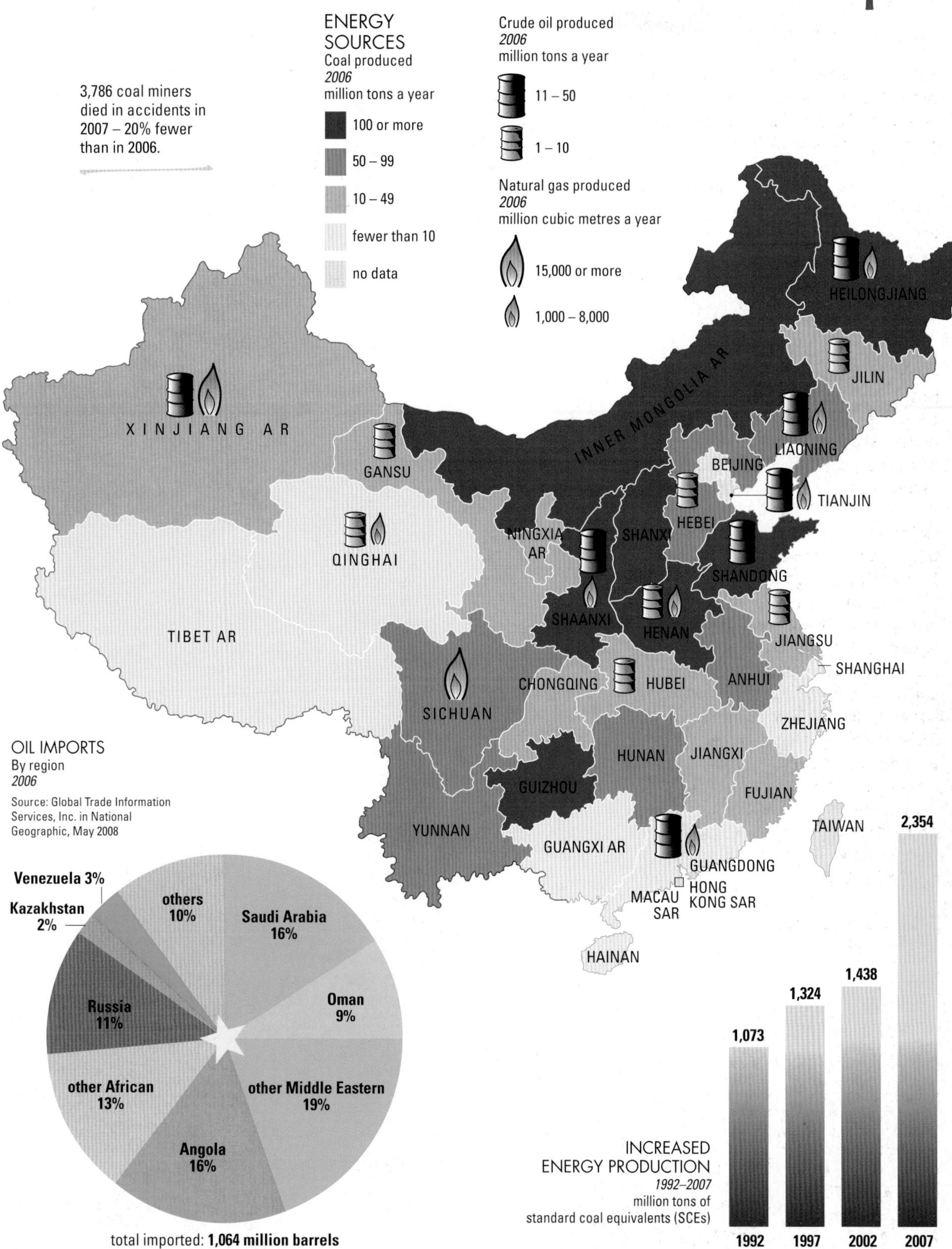

Data source: *China Statistical Yearbook 2007* and *2008* unless stated otherwise

HAIDA
INSTALL ONLY IN THIS DIRECTION
HD

Over 75% of the world's e-waste is shipped to China for disposal.

Part Four

THE ENVIRONMENT

THE STORY OF CHINESE DEVELOPMENT is tied to the parallel narrative of environmental stress. This is not solely a phenomenon of the Reform era but is a long-standing feature of a highly populated landmass with a relatively limited proportion of agricultural land. The deliberate deforestation of the Maoist period exacerbated desertification in the north, and that legacy, combined with present-day farming practices that require intensive use of water in precisely the places where it is most scarce, is a huge challenge to the continued survival of cities such as Beijing.

Ideally, a growing economy supports a more efficient infrastructure, smarter and cleaner industries, and the integrated management of key resources, such as water and fuel. In practice, although these ideals are extremely important to the government ministries concerned and to individual Chinese, the low level of environmental infrastructure and the high level of industrial and agricultural impact make the situation extremely grave in many areas.

Access to clean drinking water is a measure of the quality and indeed possibility of people's lives. It defines borders and cultures, communities, regions and language groups. And water can also be used to generate power. China's neighbors to the south, Laos, Thailand and Vietnam, are concerned that China's new hydroelectric dams will damage their livelihoods and cultures. It has been alleged that current dams are in part responsible for the high levels of the 2008 floods in the region. The Chinese government's unwillingness to explore these issues through mutual scientific investigations increases suspicions. In China itself, the "north drought, south floods" pattern of extreme weather is predicted to continue.

Flooding has plagued China for many years, and the impacts of dam technology, whilst very likely to exacerbate the problems, are not the sole cause of environmental danger to the population and the land. The dam strategy is but one part of a larger pan-national developmental strategy – the Go West campaign – which, whilst designed to open up the hinterlands to regional development opportunities, nevertheless places huge stress on the environment. Commentators have noted the creation of garbage villages, eco-refugees, and micro-economies based on the salvage of waste from China and the rest of the world.

The Go West campaign is not likely to be stopped for environmental reasons, however. It is central to the government's strategy for the management of economic inequality, particularly amongst ethnic minorities. But, the problems are massive: the western reaches are logged, and under threat of deforestation, while water is diverted to heavy industries. There are regulations to protect the air and underground water sources from further pollution, but the infrastructure to implement such good intentions are not.

There is room for hope however. China is a leapfrog economy in the environmental sector, and many politicians, industrialists, scientists and ordinary consumers are aware of the problems caused by accelerated development. Their efforts should not go unremarked. Solar energy pioneers have gained ground, and the Chongming Island eco-city in Shanghai is setting a workable example for newer urban developers and local governments. In March 2007 a law was enacted through the Ministry for Information Industries to control e-waste. It is fair to say that the political will is there, but that the pressure of development and the government's lack of transparency on environmental issues collude against the intelligence of many Chinese environmentalists.

MONEY CAN MOVE EVEN THE GODS

China is undergoing the largest internal migration in the history of the world. 120 million people have moved to the cities, and 80 million to small towns, seeking work.

Within the next decade, China will be transformed from a predominately rural society into an urban one, and by 2020 at least 60 percent of people are expected to be living in cities and towns.

Farmers have moved off the land in search of a better living. Rural migrants (*mingong*) form the bulk of the labor force for the massive construction projects taking place in the cities, most spectacularly for the Beijing Olympics in 2008 and the Shanghai International Exposition in 2010. They work long hours, receive low wages and live in cramped and basic conditions, but manage to save money to send home to their villages. In this way the effects of the east-coast economic boom slowly trickle through to the rural areas.

The soaring urban population has created resource and environmental problems. The system of registration that determines someone's official place of residence (*hekou*) has made it difficult for rural migrants to claim housing and welfare benefits. The controls are being relaxed, however, and protective labor legislation enacted. Migrants also find their own solutions. For example, migrants have set up schools in Beijing for an estimated 300,000 to 500,000 children.

▸▸ *see also page 114*

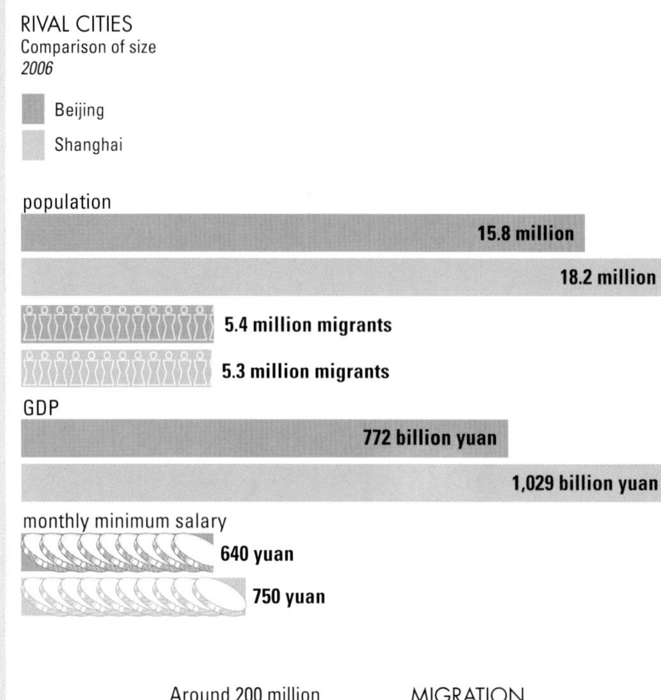

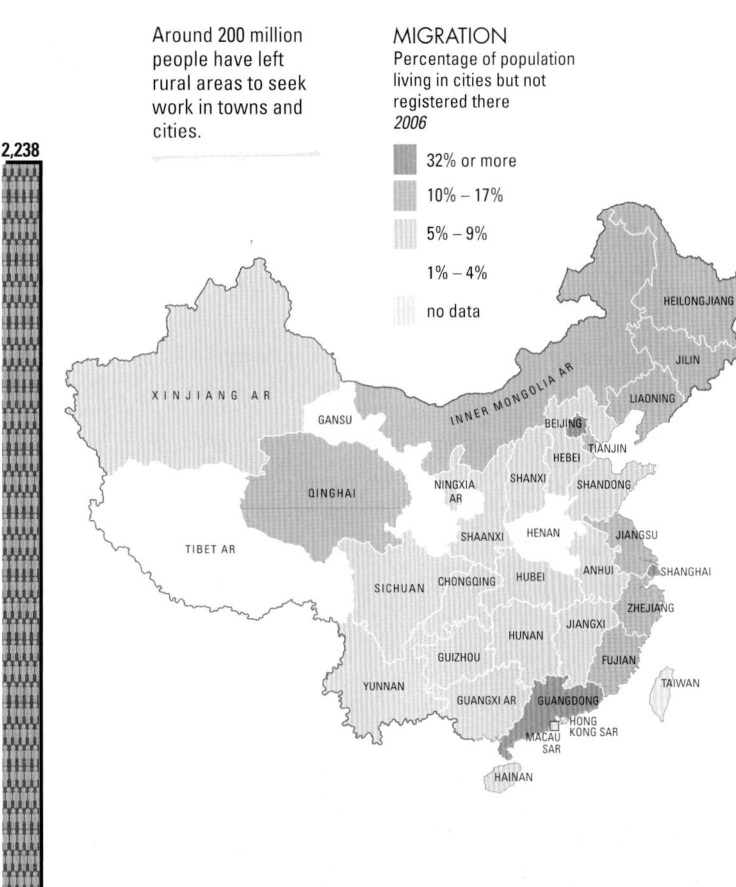

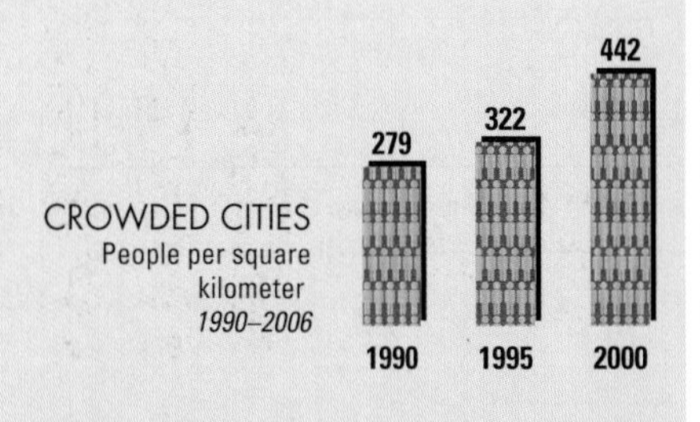

 Copyright © Myriad Editions

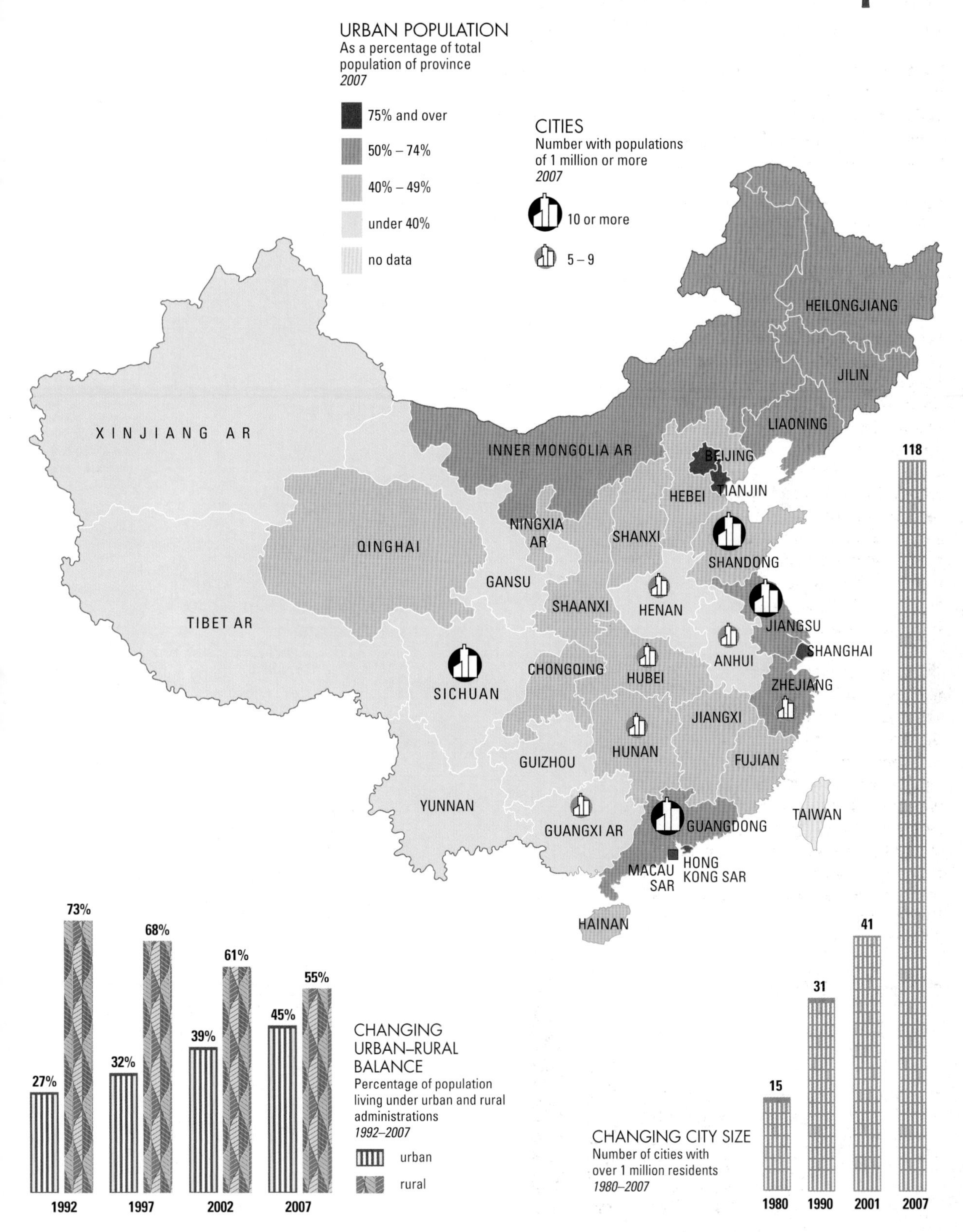
URBAN POPULATION
As a percentage of total population of province
2007
75% and over
50% – 74%
40% – 49%
under 40%
no data
CITIES
Number with populations of 1 million or more
2007
10 or more
5 – 9
XINJIANG AR
TIBET AR
QINGHAI
GANSU
NINGXIA AR
INNER MONGOLIA AR
HEILONGJIANG
JILIN
LIAONING
BEIJING
TIANJIN
HEBEI
SHANXI
SHANDONG
SHAANXI
HENAN
JIANGSU
SHANGHAI
ANHUI
SICHUAN
CHONGQING
HUBEI
ZHEJIANG
JIANGXI
HUNAN
FUJIAN
GUIZHOU
YUNNAN
GUANGXI AR
GUANGDONG
TAIWAN
MACAU SAR
HONG KONG SAR
HAINAN
CHANGING URBAN–RURAL BALANCE
Percentage of population living under urban and rural administrations
1992–2007
urban
rural
27%
73%
1992
32%
68%
1997
39%
61%
2002
45%
55%
2007
CHANGING CITY SIZE
Number of cities with over 1 million residents
1980–2007
15
1980
31
1990
41
2001
118
2007

JOURNEY TO THE WEST: FOUR WHEELS GOOD, TWO WHEELS BAD

The growth of motorized traffic in China's cities – especially in those along the eastern seaboard – is both indicative of the drive towards rapid modernization, and of the dangers that development can produce.

Up to 90 percent of emissions in major cities can be attributed to motor vehicles, car accidents are common, and luxury car ownership is on the rise. China produced 8.9 million motor vehicles in 2007, and private ownership of cars increased 27 percent on the previous year. In Beijing alone, 1,000 new cars are rolling on to the streets every day.

Air travel is also becoming increasingly popular, and is more frequently used as a means of crossing China's vast area. Domestic passengers make up more than 90 percent of the country's air travelers.

see also page 115

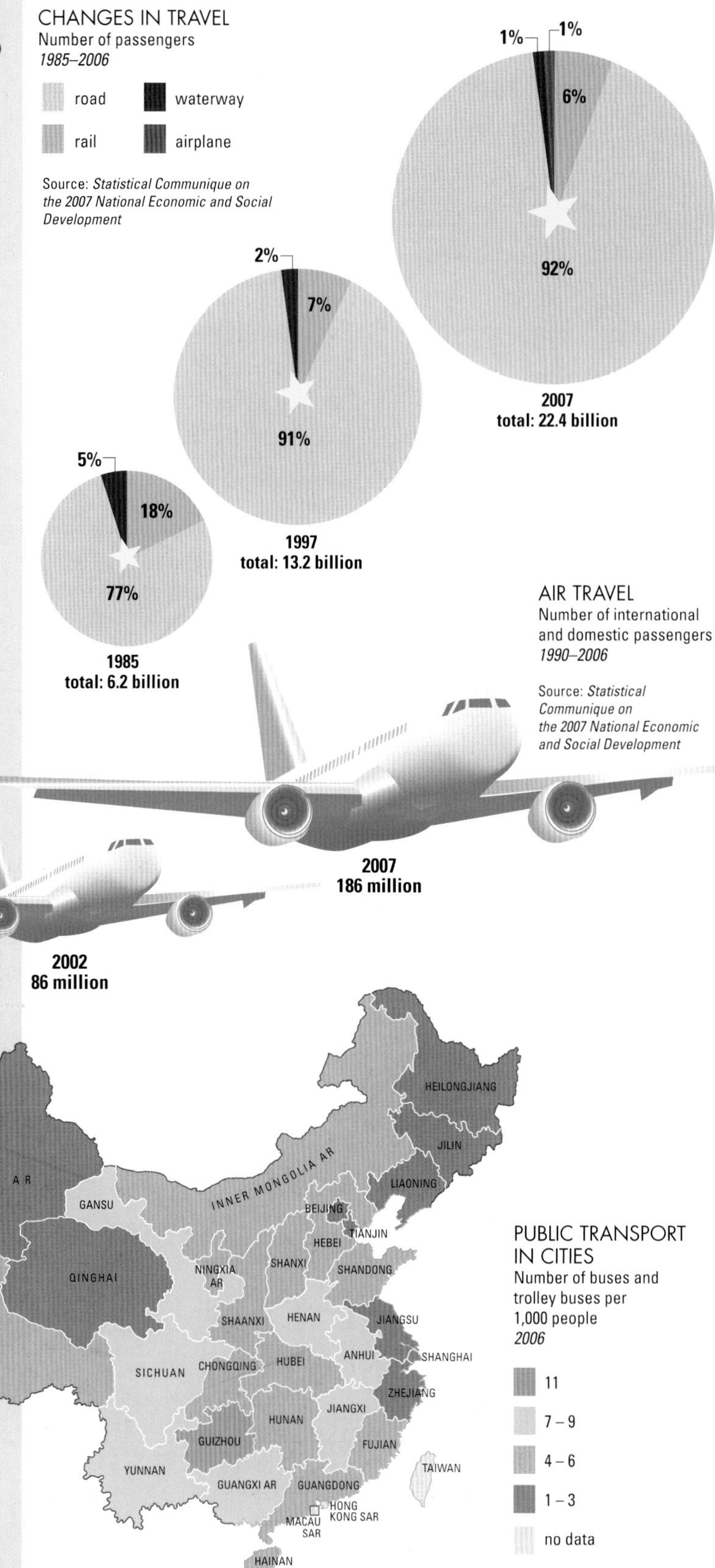

 Copyright © Myriad Editions

A massive road-building program doubled the length of China's roads between 2001 and 2006.

Despite the significant increase in the number of private cars, the number of people killed in traffic accidents is decreasing.

The number of passenger vehicles on China's roads doubled between 2004 and 2007.

Data source: *China Statistical Yearbook 2007* and *2008* unless stated otherwise

THE SKY IS BLUE, THE WATER IS CLEAR, AND BEIJING IS BECOMING MORE AND MORE BEAUTIFUL

One of China's foremost dilemmas is the conflict between environmental protection and economic growth.

Air pollution has increased in line with industrial expansion and the massive rise in private car ownership, and has led to smog being a feature of many of China's cities. In preparation for the 2008 Beijing Olympics, industry was either moved out of the city or shut down, and restrictions were placed on vehicle use in an attempt to improve the air quality.

China's State Environmental Protection Administration (SEPA) is trying to tackle the problem of pollution with a number of innovative policies, but it has had limited success in rolling these out across the country. Only around 10 percent of environmental laws and regulations are actually enforced. Public opposition to pollution seems to be growing, however, with some high-profile protests, and thousands of smaller ones, being held against air and water pollution.

Carbon dioxide emissions, being invisible, receive less attention in China, but in 2007 the country gained the dubious distinction of overtaking the USA as the world's largest emitter of this greenhouse gas. Its emissions per capita are still, however, relatively modest.

▸▸ *see also page 115*

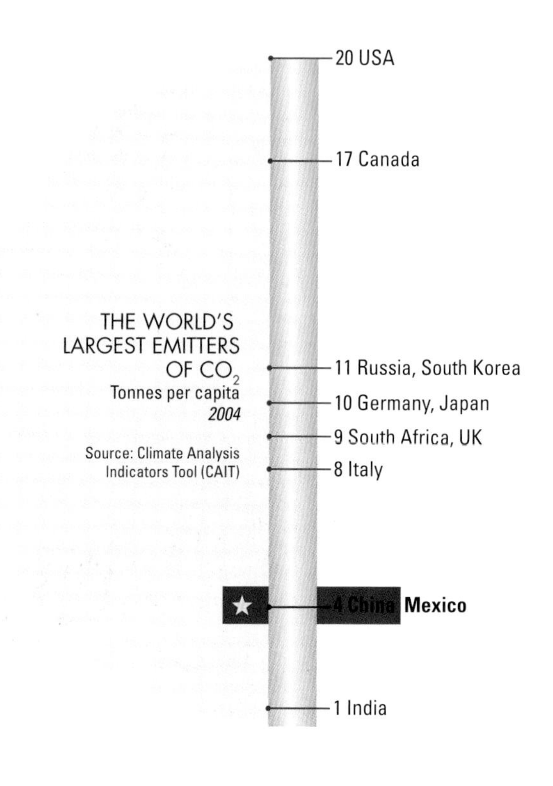

THE WORLD'S LARGEST EMITTERS OF CO_2
Tonnes per capita
2004

Source: Climate Analysis Indicators Tool (CAIT)

In 2001, China pledged to reduce emissions of sulfur dioxide by 10% by 2005. They rose by 27%.

Source: *Foreign Affairs* Sept/Oct 2007

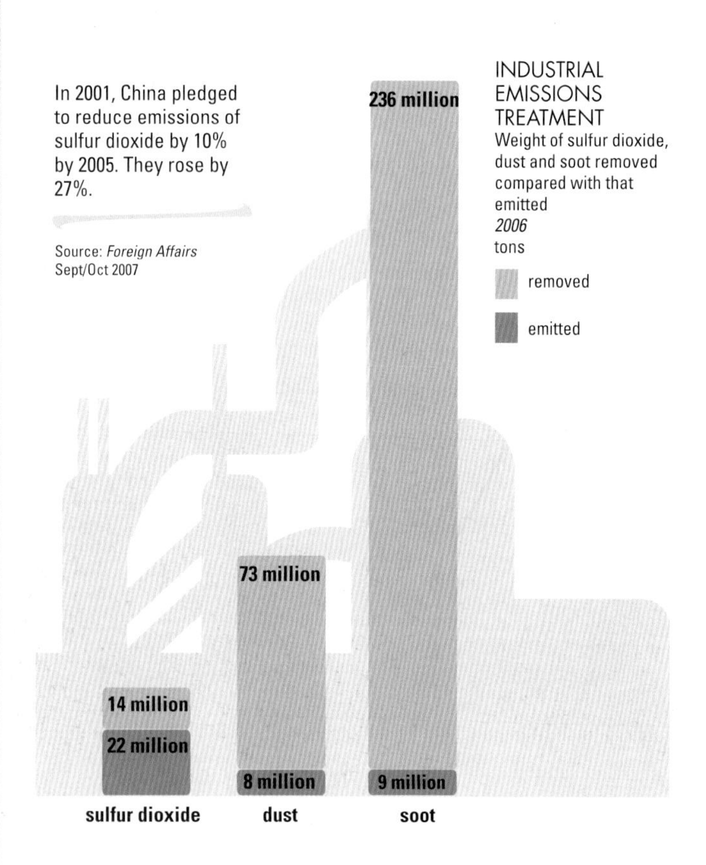

INDUSTRIAL EMISSIONS TREATMENT
Weight of sulfur dioxide, dust and soot removed compared with that emitted
2006
tons

 Copyright © Myriad Editions

AIR POLLUTION 中 20

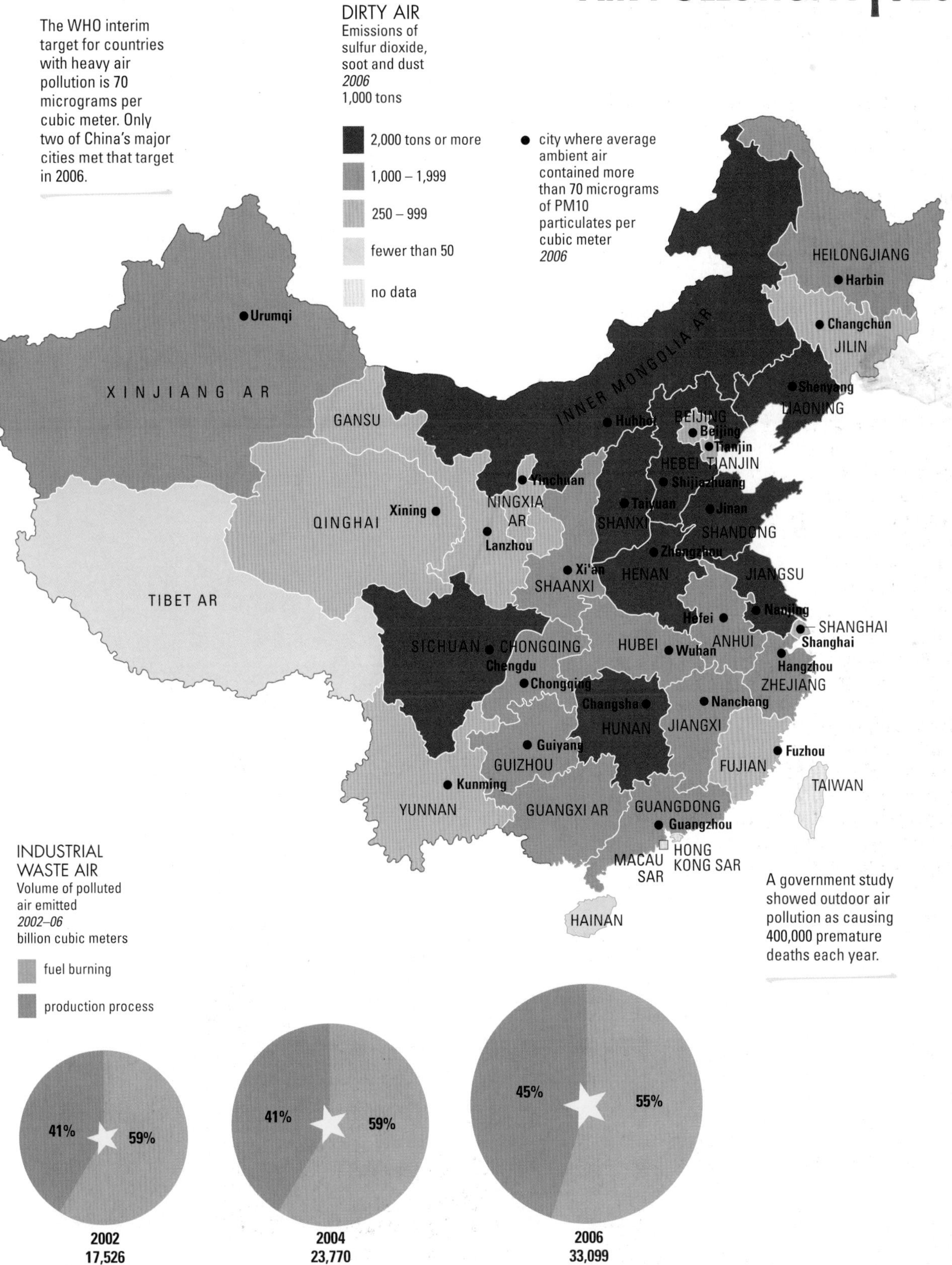

NOT SINKING A WELL UNTIL ONE IS THIRSTY

China is facing a water crisis – both in terms of supply and of quality.

In the northeast, where half the country's population rely on just 15 percent of the country's water, both industry and agriculture are placing huge demands on available resources, using up groundwater at an unsustainable rate. The water table is dropping rapidly, and many small farmers are finding it hard or impossible to dig deep enough wells. Schemes are underway to divert water from the south to the north, but this will provide only a small fraction of the shortfall and may well cause problems in the more rural south.

A more long-term problem will arise if the glaciers in the mountainous southwest, which feed many of China's rivers, continue to retreat at current rates (50 square miles a year). Loss of this massive natural store will severely affect the country's water supply at a time when its increasing population is already predicted to lead to water stress.

China's rivers and lakes are becoming increasingly polluted from industrial effluent, which is killing fish, contaminating drinking water, and leading to serious health problems for anybody who continues to try and use the water. This is a particular problem in rural areas, where an estimated 320 million people do not have access to clean drinking water.

▸▸ *see also page 116*

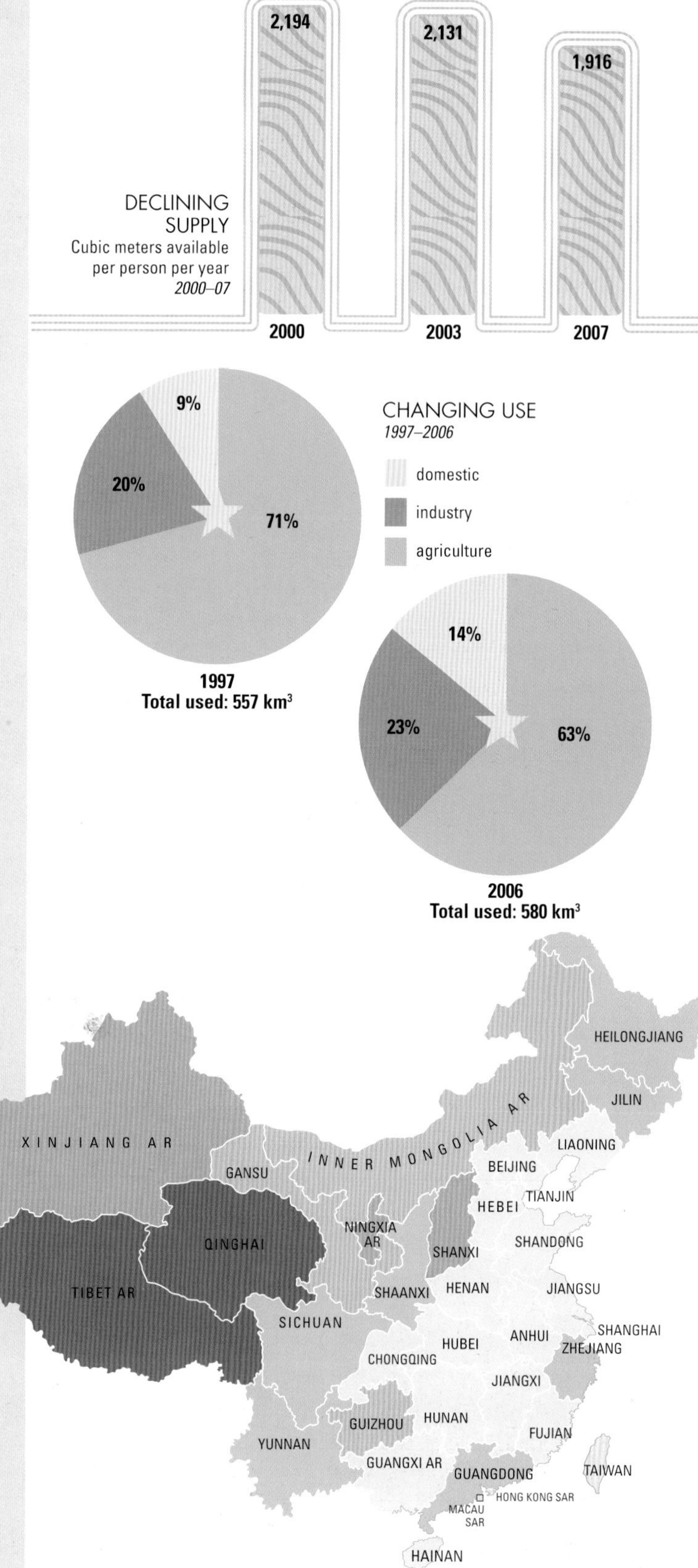

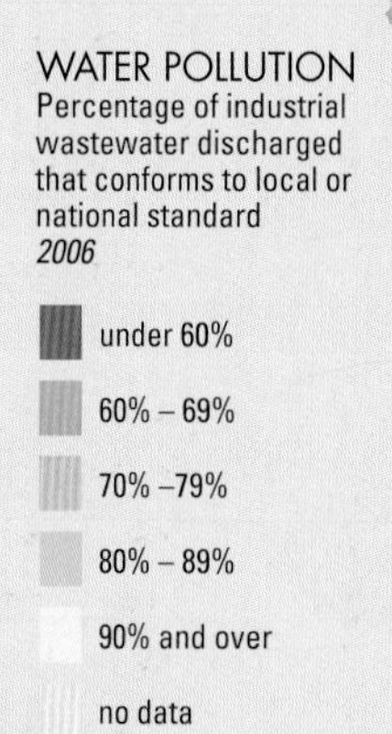

 Copyright © Myriad Editions

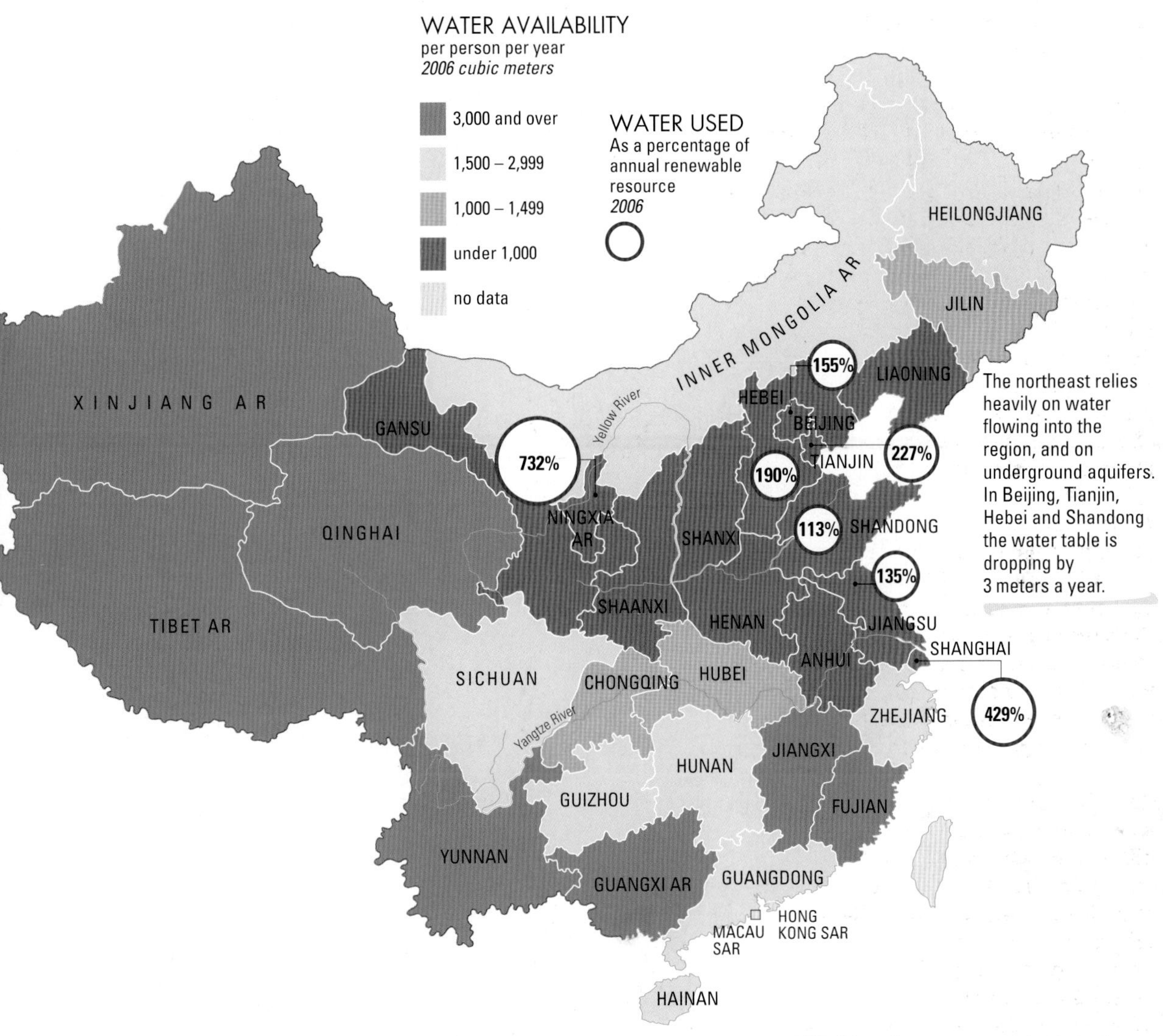

The northeast relies heavily on water flowing into the region, and on underground aquifers. In Beijing, Tianjin, Hebei and Shandong the water table is dropping by 3 meters a year.

Work is underway on a grand scheme to supplement the Yellow River with water from the Yangtze. The eastern route was projected to be finished in 2008, the middle route, which includes 1,200 km of canals, is projected for completion in 2030, and the high-altitude western route in 2050.

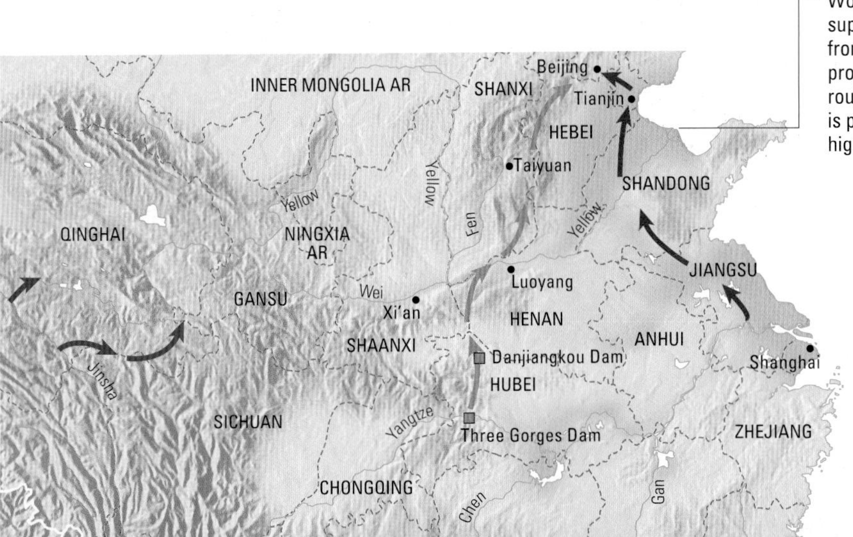

WATER FOR THE NORTH

- river
- provincial boundary

South-to-North water diversion:

- eastern route
- middle route
- western routes

Source: www.water-technology.net

Data source: *China Statistical Yearbook 2007* and *2008* unless stated otherwise

There were 90,000 recorded protests in 2006 – nine times more than in 1994.

Part Five
GOVERNANCE AND RIGHTS

CHINA'S POLITICAL REFORMS have not kept pace with the rapid restructuring of the economy. The Communist Party retains it monopoly of power, including its control of the military, and sustains an authoritarian state, with its repression and human rights abuses. Nonetheless, important changes are shaping China's political development.

Firstly, the gap between the economic reforms and the political reforms are deliberate policy. Moving away from a planned and command economy towards a market-led economy is reforming a revolution. It is no wonder that the Party-State's ruling elite insist that political stability is equal in priority to the market reforms. Even the call for "social harmony" has been played down in favor of "scientific development." It is important to remember, however, that political stability is valued in its own right, given the history of modern China's turbulence and disorder.

Second, the record of political reform is mixed. The bureaucracy has been modernized and is increasingly professional in terms of education, training, and the competence of its officials. The Party has also attempted to broaden its representation in line with Jiang Zemin's "Three Represents", which called for the admission of entrepreneurs and more intellectuals to Party membership. But power remains firmly in the hands of a small elite. This is evidenced by the unopposed succession to office of the Fourth Generation of Party-State leaders, their consolidation of power, and the gradual withdrawal from public life of the former Party-State chief, Jiang Zemin, as Chairman of the Central Military Commission, and his replacement by the current Party Secretary and State President, Hu Jintao. A fifth generation of leaders is already being carefully groomed for a smooth takeover in 2012.

The 17th Party Congress in October 2008 trumpeted improvements in rural self-rule and the democratic rights of farmers, but there is little evidence of progress as yet. While villages have been electing leaders and holding them to account since 1988, this has not spread to the higher levels of towns and counties. In the cities, the Community Residents' Committees have greater responsibilities than they did in the 1950s, particularly in regard to the provision of services, and are electing younger, better-educated leaders. In addition, residents' housing associations have been active in protecting and furthering their interests. Although this grassroots activity bodes well for participatory practice, much will depend on the extent of Party control.

Thirdly, with the rapid growth of the market-led economy, producers and consumers have formed organizations to represent, promote and defend their interests. In China, these are mainly trade and professional associations and they are obliged to register with the government for legal recognition. In addition, there are the Communist Party pillar or mass organizations, government-sponsored and independent NGOs, the official and underground organizations. Also growing in numbers, are the often spontaneous, and usually local, protest movements, organized by farmers, workers pensioners and ethnic minorities, for example. They do not threaten the regime but they affect public order and erode political stability.

Another view is that political movements, along with the organized interests, promote and enrich civil society. Whether or not the development of civil society in an authoritarian Party-State is a harbinger of radical change, it is an important arena for political activity, conflict and participation.

THE PEOPLE ARE THE WATER, THE RULER A BOAT. THE WATER CAN KEEP THE BOAT AFLOAT. THE WATER CAN ALSO CAPSIZE IT.

China is a one-party state. Supreme power is exercised by the nine members of the Standing Committee of the Politburo of the Communist Party.

The Politburo's Standing Committee is ultimately responsible for maintaining the Party's monopoly of power. The nine members ensure this by occupying the commanding positions in the State Council or government, the National People's Congress or legislature, and in the Military Commission or armed forces.

They rule by the authority vested in them by the Party. The National Party Congress meets once every five years to elect the Central Committee, currently 204 members, which in turn elects the 25 members of the decision-making Politburo.

▸▸ *see also page 116*

LEGISLATURE

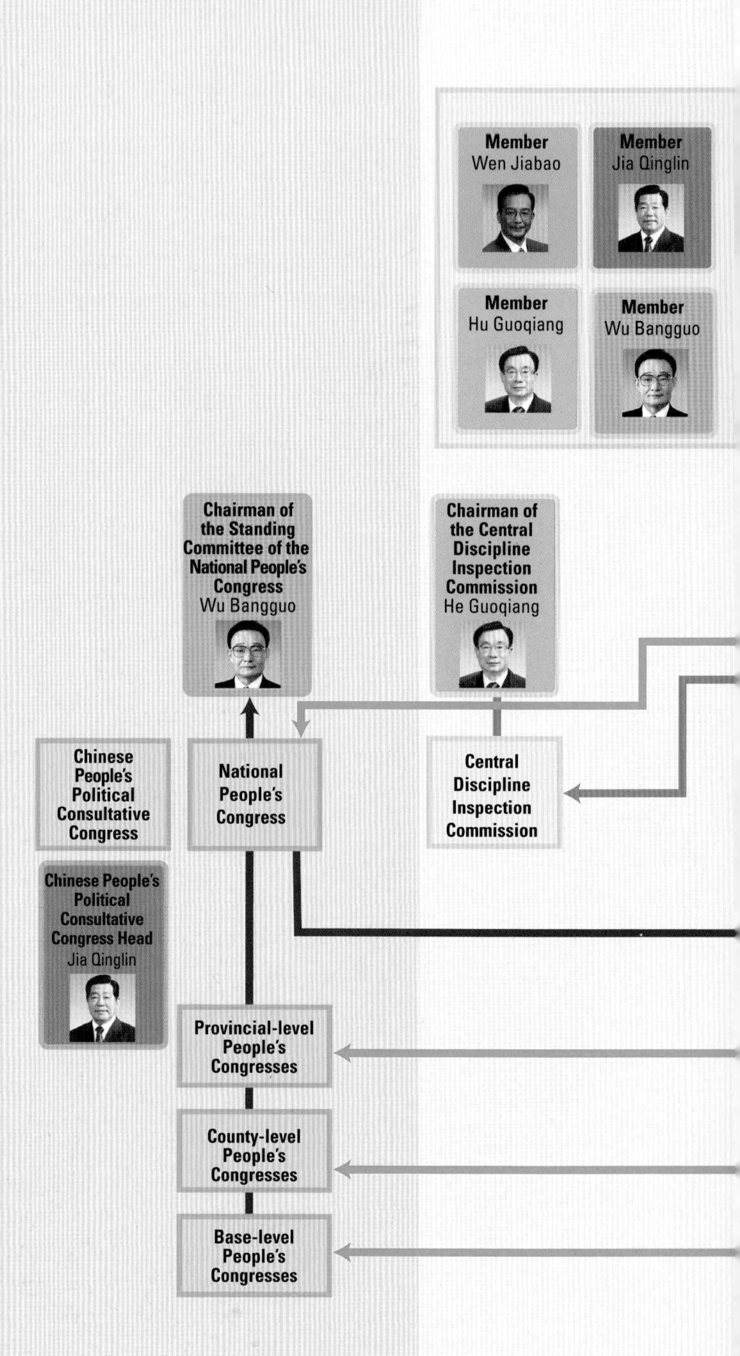

POWER STRUCTURE OF CENTRAL GOVERNMENT
2008

Source: author, Blecher

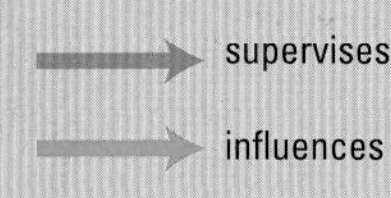

 Copyright © Myriad Editions

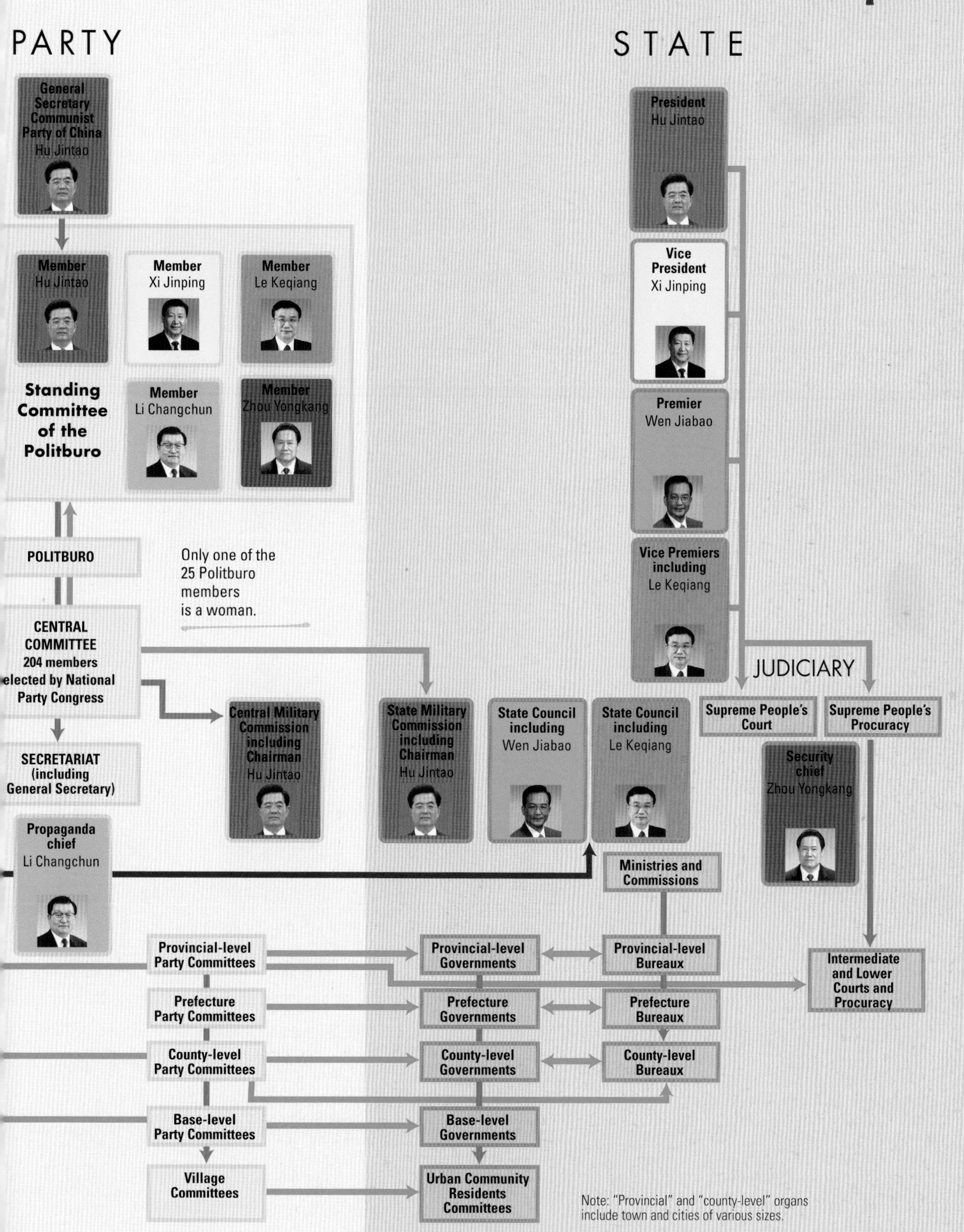

Note: "Provincial" and "county-level" organs include town and cities of various sizes.

A TIGER WHOSE BUTTOCKS MAY NOT BE TOUCHED

The Chinese Communist Party (CCP) is the most powerful political party in the world.

It maintains this power even though fewer than 6 percent of people are members. Many more apply to join and fail to pass the rigorous selection procedure.

Members of the party are continually trained and educated in the party principles and processes through the Party Schools, which exist at all levels of the Party. The Central Party School in Beijing is important for providing opportunities for networking, as well as serving as a platform for announcement of important policies and decisions.

The National Party Congress takes place every five years, and is attended by over 2,000 delegates, elected through the party hierarchy. The Party claims that the composition of the 17th Party Congress in the autumn of 2007 was its most representative, with 30 percent elected from the grassroots, but it was still the case that over 90 percent of delegates had received higher education.

The eight small democratic parties under the leadership of the CCP provide expertise and serve in an advisory capacity to the Party-State. One of their characteristics is the relatively high proportion of women, when compared with the significantly lower proportion in the decision-making bodies of the Party.

▶▶ *see also page 117*

INCREASE IN PARTY MEMBERSHIP

Number of members of the CCP at time of each National Party Congress

Sources: Saich, 2004; www.chinatoday.com

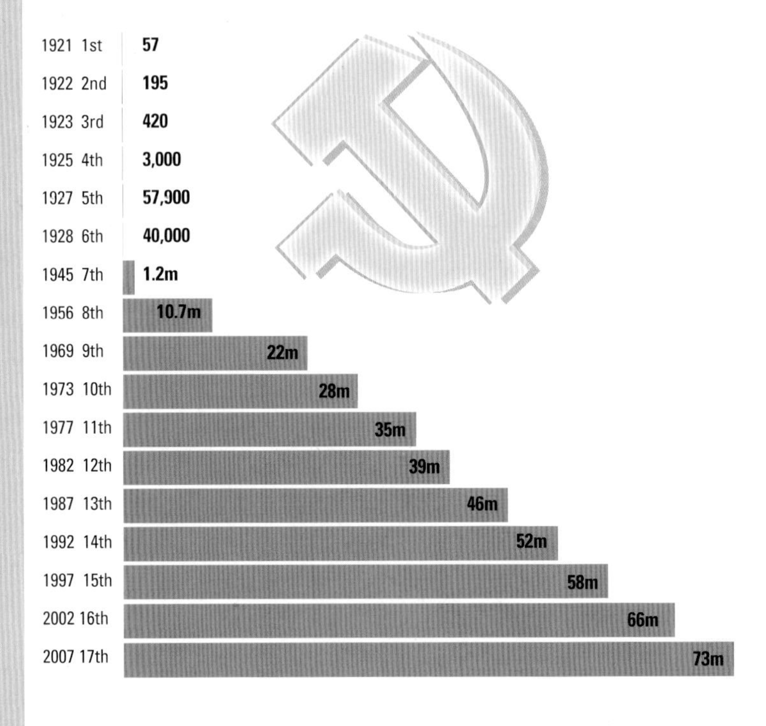

MEMBERSHIP

of Chinese Communist Party and of its Central Committee

Source: *Women and Men in China*

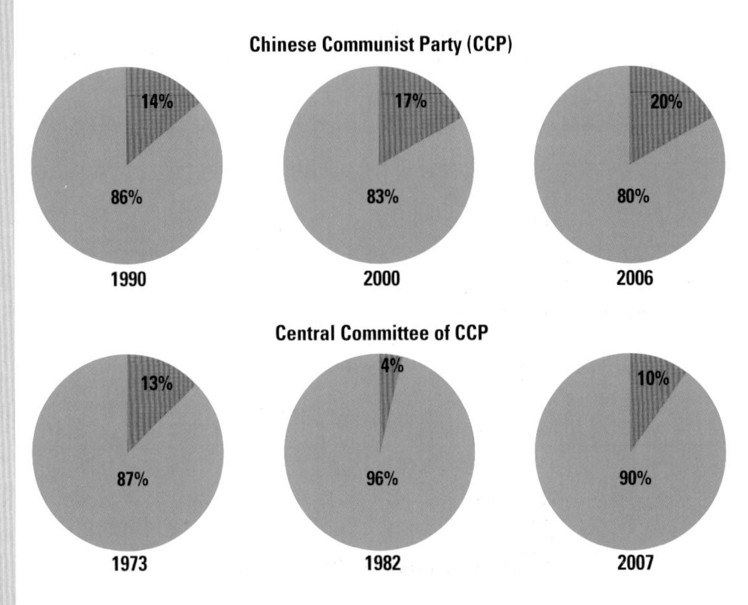

 Copyright © Myriad Editions

At the Party center are eight more-or-less permanent groups, known as the Leading Groups. They report directly to the Politburo Standing Committee and to the Party Secretariat, and are headed by a member of the Standing Committee.

Although they are informal, it is arguable that these groups should be counted as among the most powerful bodies in China. They are not reported in the media and do not appear on organizational charts.

"DEMOCRATIC" PARTIES
Under the leadership of the Chinese Communist Party
2006

women
men

Source: www.cppcc.gov.cn

Taiwan Democratic Self-Government League
Membership: those of Taiwan origin working for peaceful reunification
2,000

China Zhi Gong Dang
Membership: returned overseas Chinese families
2,800

Revolutionary Committee of the Guomindang
Membership: former members and their families engaged in poverty alleviation, medical care and establishing schools
34% 66% 82,000

Chinese Peasants' and Workers' Democratic Party
Membership: those engaged in medicine, health, science and technology
49% 51% 100,000

China Association for Promoting Democracy
Membership: those engaged in education, culture and publishing
47% 53% 103,000

September 3rd Study Society
Membership: those involved in science and technology
36% 64% 105,000

China Democratic National Construction Association
Membership: those engaged in business, research and government
31% 69% 108,000

China Democratic League
Membership: academics in the natural and social sciences
38% 62% 182,000

SHARPENING THE WEAPONS AND FEEDING THE HORSES

The PLA is controlled by the Party-State through the two arms of the Central Military Commission, both of them chaired by Party-State boss Hu Jintao.

The relationship between the Party-State and the PLA is a symbiotic one, however, for the Party-State relies ultimately on the PLA for maintaining domestic stability and for helping to project China as an international power, while the PLA is beholden to the Party-State for its large-scale modernization program.

The substantial increases in the military budget over the last 15 years testify to this support. The immediate focus for the military build-up, however, is only 160 km across the Taiwan Straits. Even so, this may have less to do with national security or power than with national pride.

▸▸ *see also page 118*

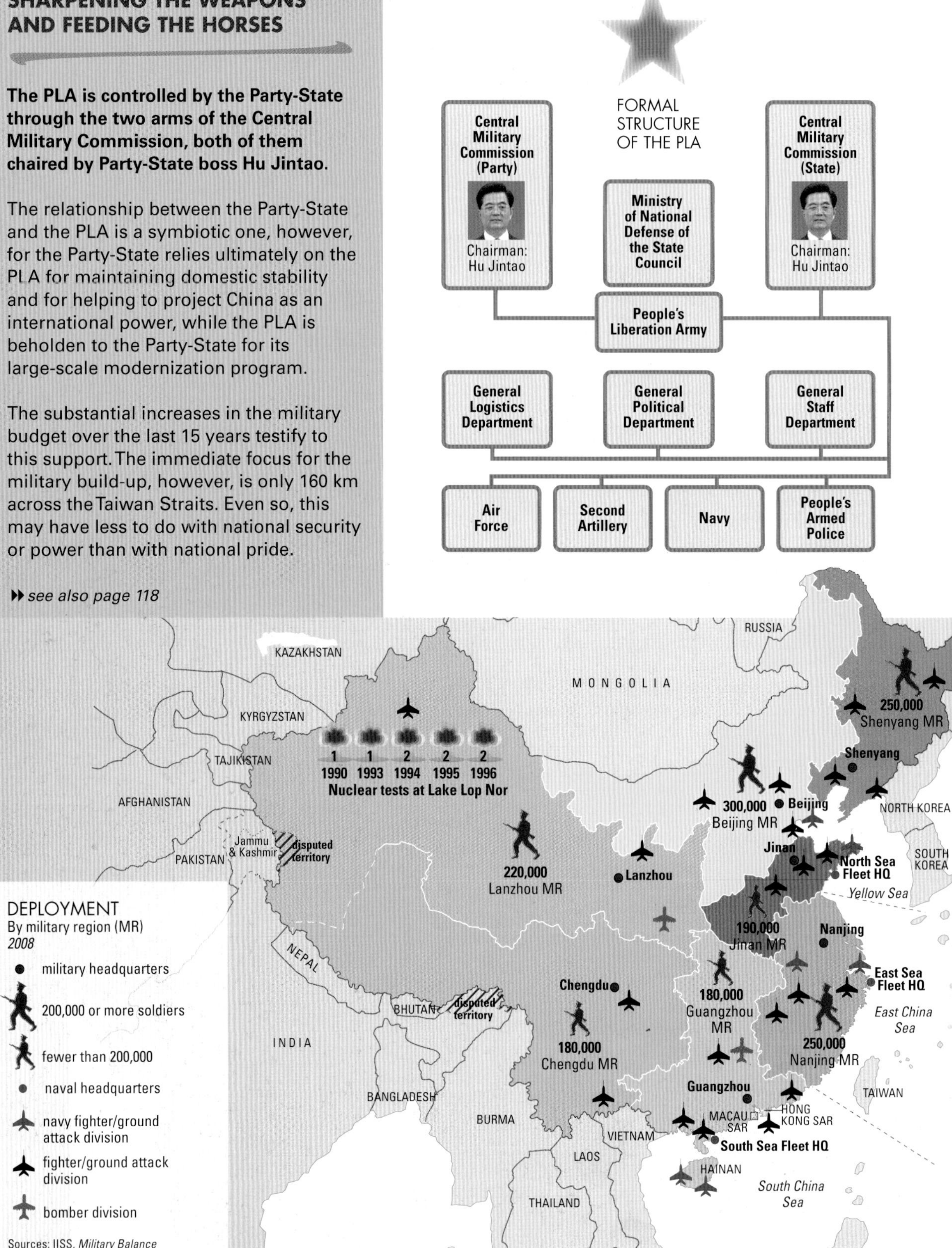

 Copyright © Myriad Editions

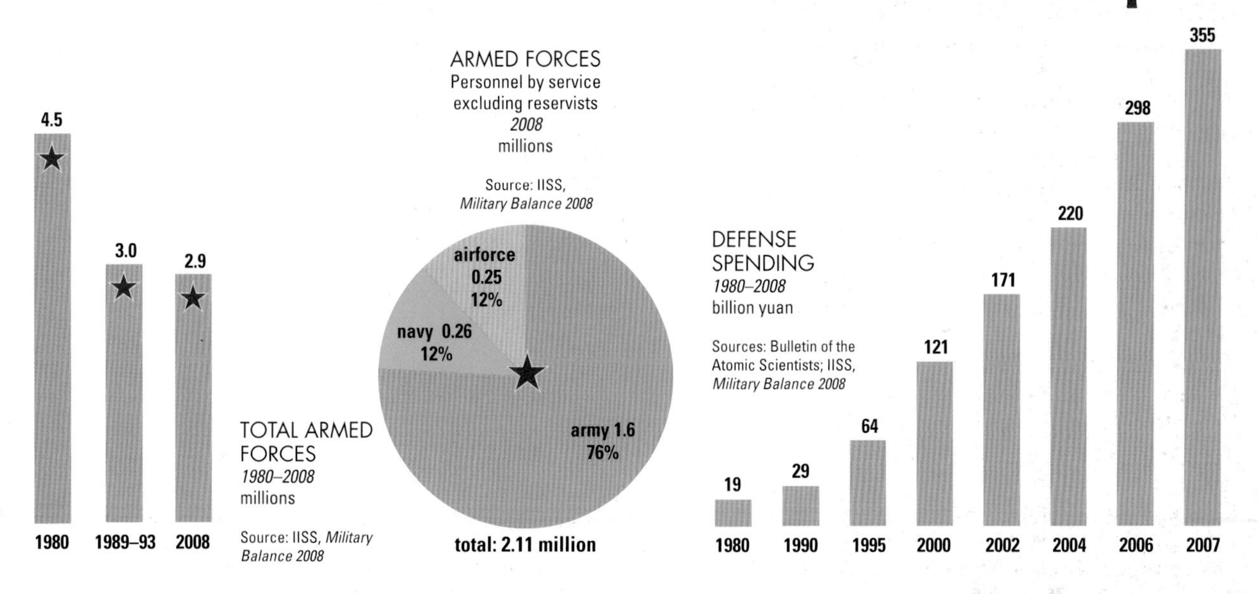
TOTAL ARMED FORCES
1980–2008
millions
4.5
3.0
2.9
1980
1989–93
2008
Source: IISS, Military Balance 2008
ARMED FORCES
Personnel by service excluding reservists
2008
millions
Source: IISS, Military Balance 2008
airforce 0.25 12%
navy 0.26 12%
army 1.6 76%
total: 2.11 million
DEFENSE SPENDING
1980–2008
billion yuan
Sources: Bulletin of the Atomic Scientists; IISS, Military Balance 2008
19
29
64
121
171
220
298
355
1980
1990
1995
2000
2002
2004
2006
2007

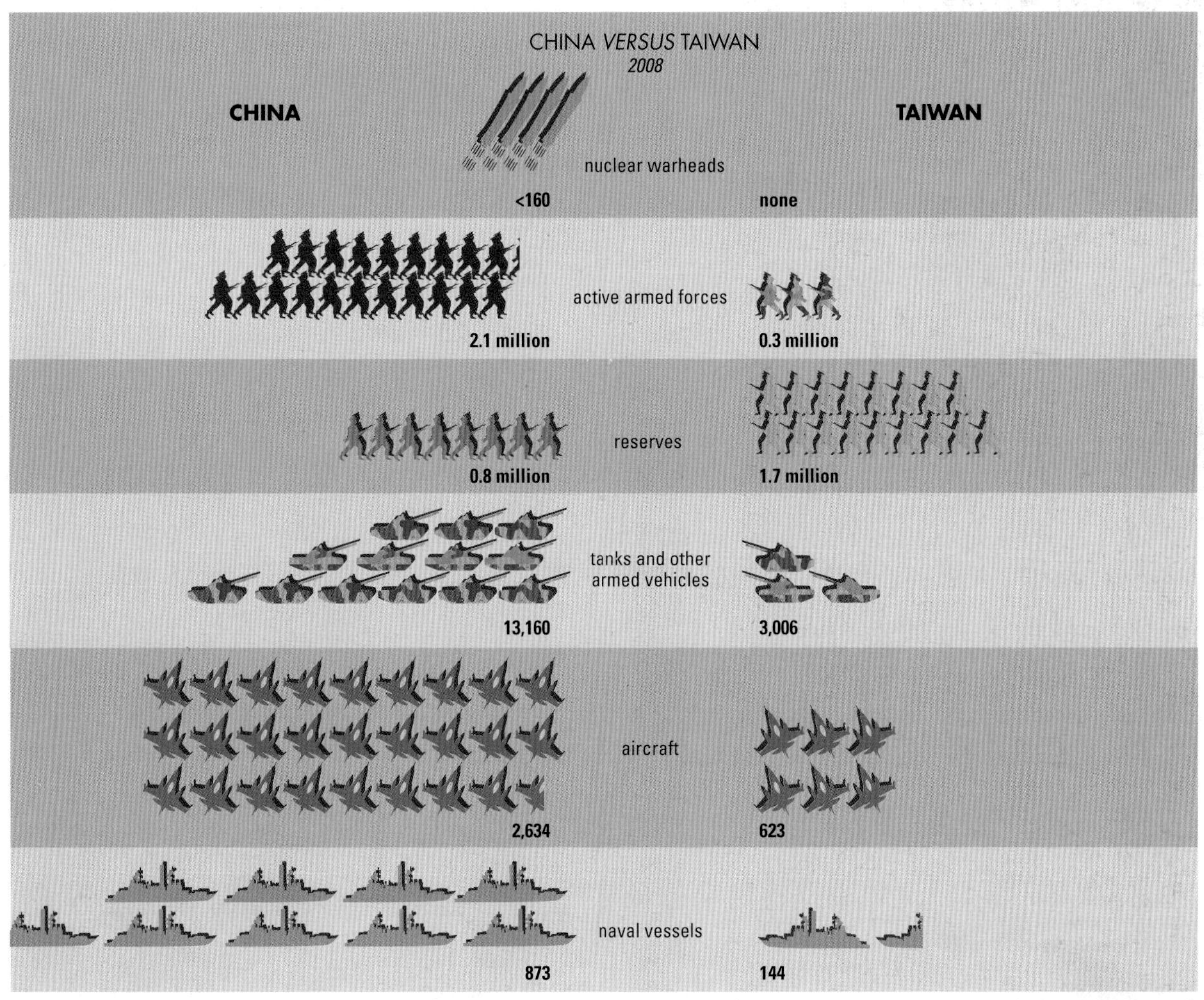
CHINA VERSUS TAIWAN
2008
CHINA
TAIWAN
nuclear warheads
<160
none
active armed forces
2.1 million
0.3 million
reserves
0.8 million
1.7 million
tanks and other armed vehicles
13,160
3,006
aircraft
2,634
623
naval vessels
873
144

MOUNTAINS AND RIVERS ARE EASY TO MOVE BUT IT IS IMPOSSIBLE TO CHANGE THE NATURE OF A MAN

In the era of economic reform, China has taken an important step towards legality, moving away from the Maoist principle of "rule by persons" to that of "rule by law" and towards enshrining the rule of law in the constitution and practice.

The weight of the Party however, still looms over the legal system. Although citizens are increasingly turning to the courts to settle disputes, and the administration of civil and commercial law has reached a good professional standard, criminal justice is another matter. The accused have often been denied the rights to which they are entitled, and defense lawyers can be discriminated against and subject to harassment.

The Party-State is preoccupied with maintaining economic and political stability. There have been a number of high-profile cases of corruption since the market reforms, the most recent being the conviction of a former vice-mayor of Beijing for taking bribes related to construction work for the Olympic Games. Social harmony is being enforced at whatever the cost. It is not uncommon for protests and demonstrations to be violently suppressed and petitioners brutally treated.

▸▸ *see also page 118*

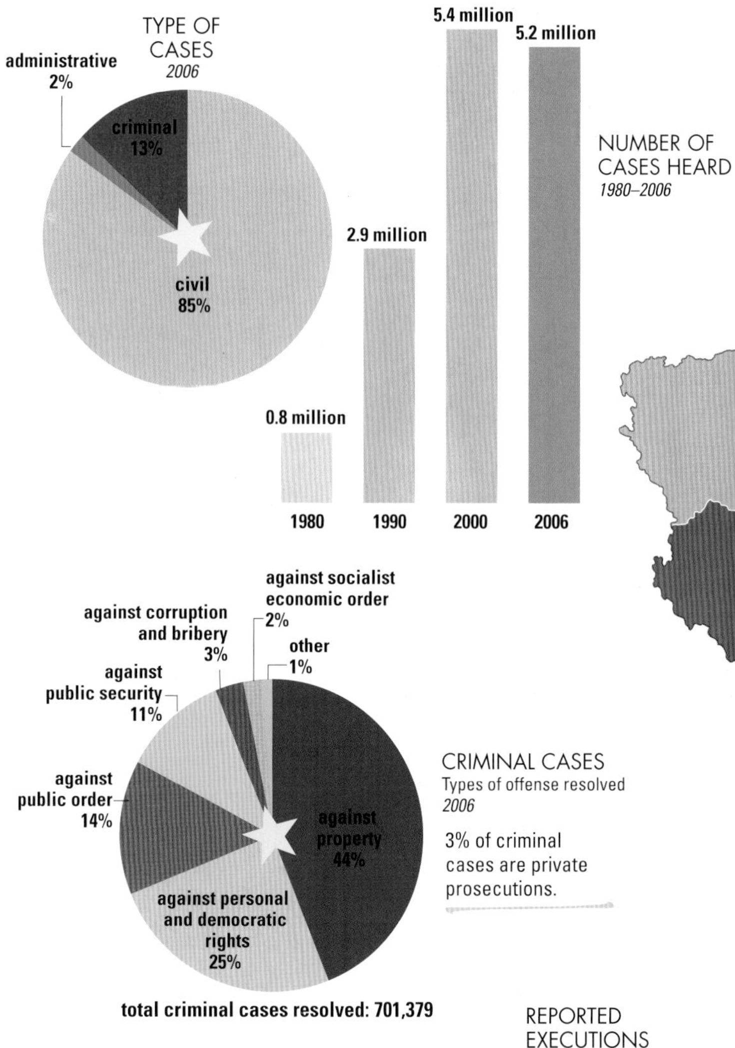

DEATH PENALTY

The number of executions carried out annually in China is unknown. Amnesty International records those listed in public reports, but stresses they these are only the tip of the iceberg. In March 2004, a senior delegate from Chongqing Municipality estimated the figure at nearly 10,000. Amnesty International estimated 8,000 executions took place in 2006, and a US-based advocacy group, Dui Hua Foundation, documented 6,000 in 2007. Since early in 2006, the Supreme Court has been required to give approval to death penalties and, according to official sources, has overturned around 15 percent of them.

The death penalty can be given for 68 offences, two-thirds of which are non-violent crimes such as bigamy, internet-hacking and cyber crimes, stealing petrol and tax evasion. In 2008, the execution of eight drug dealers was used to draw attention to the government's crackdown on drugs trafficking.

REPORTED EXECUTIONS IN CHINA
As recorded by Amnesty International
2004–07

Source: Amnesty International USA

3,400+
1,760+
1,010+
470+
2004
2005
2006
2007

REPORTED EXECUTIONS WORLDWIDE
As recorded by Amnesty International
2007

Source: Amnesty International USA

China 470+
Iran 317+
Saudi Arabia 143+
Pakistan 135+
USA 42
Iraq 33+
Vietnam 25+
other 87

 Copyright © Myriad Editions

LAW ENFORCEMENT EXPENDITURE

Annual spending on public security and armed police per person
2002
yuan

China average: 162 yuan

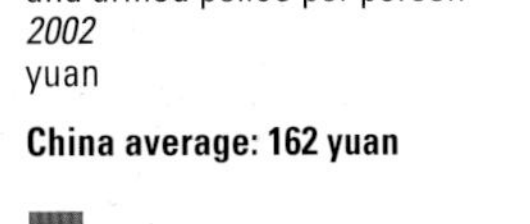

In 2002, China had one lawyer for every 12,700 citizens, but by 2006 this ratio had improved to one for every 10,770 citizens. Even so, the majority of criminal defendants have no legal representation.

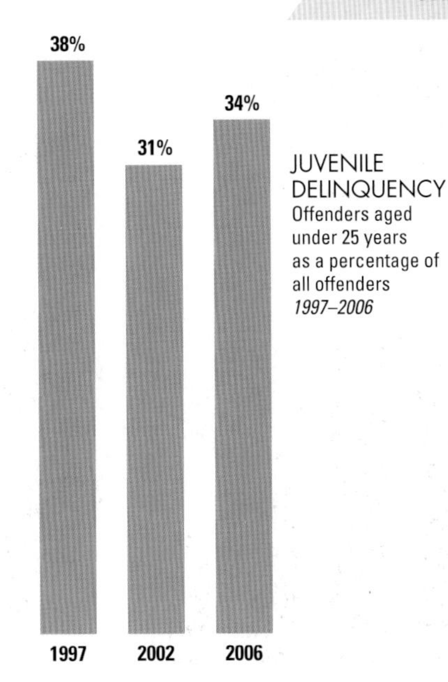

JUVENILE DELINQUENCY

Offenders aged under 25 years as a percentage of all offenders
1997–2006

LEGAL SYSTEM

Source: Chinese Academy of Social Sciences, 1989

CONSTITUTION
modified by the National People's Congress

FUNDAMENTAL LAWS
enacted and modified by the National People's Congress

LAWS
enacted and modified by the Standing Committee of the National People's Congress

ADMINISTRATIVE RULES AND REGULATIONS
enacted and revised by the State Council

REGULATIONS
made by Ministries or Committees of the State Council

LOCAL RULES AND REGULATIONS
enacted and revised by the People's Congress and its Standing Committee of Provinces, Autonomous Regions or Municipalities directly under the central government

RULES AND REGULATIONS
made by the People's Government of Provinces, Autonomous Regions and Municipalities directly under the central government

Data source: *China Statistical Yearbook 2007* unless stated otherwise

> **"USE POWER FOR THE PEOPLE, SHOW COMPASSION FOR THE PEOPLE AND SEEK BENEFIT FOR THE PEOPLE."**
>
> **Hu Jintao**

China is a more open society than 30 years ago.

Personal freedoms have expanded through the extension of property rights, freedom of movement, technological innovations and a more tolerant attitude towards expressions of political opinions. The growing number of social organizations, 168,000 in a recent count, suggests that civil society may be taking root.

Although these organizations must be officially registered, they include advocacy groups, like those that successfully campaigned for the improvement of disablement rights. The record however is mixed. The contributions of charities during the Sichuan province earthquakes must be balanced against the repression of dissidents and the suppression of ethnic minorities in Tibet and Xinjiang.

▸▸ *see also page 118*

PROTESTS AND RIOTS

Opposing land appropriation, corruption and environmental issues

1994–2006

Source: *Economist*

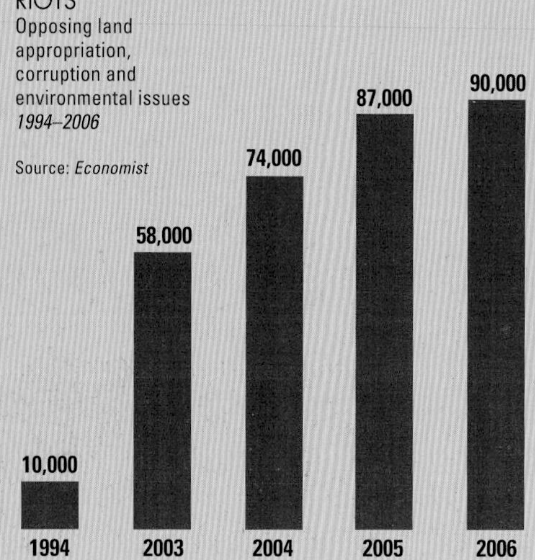

INTELLECTUAL FREEDOM

1957–2008

action encouraging freedom

action discouraging freedom

Year	Event	Action
1957	Chinese intellectuals, encouraged by Mao to criticize Communist Party, attack its right to govern.	encouraging
1957–58	Crackdown on intellectuals.	discouraging
1966–76	Cultural Revolution brings further suppression of intellectuals.	discouraging
1978	Democracy Wall set up in Beijing, on which citizens are encouraged to paste political tracts.	encouraging
1979	Wall closed.	discouraging
1989	Tiananmen Square demonstration for democratic reform is brutally suppressed.	discouraging
1997	A sexual rights movement is established	encouraging
1999	Falun Gong banned under law against cults.	discouraging
2000	Party membership opened up to entrepreneurs, intellectuals and scientists.	encouraging
2002	Government promotes the internet because of its economic benefits, but exercises control over information.	discouraging
2004	Human rights guaranteed and private property protected under amendment to state constitution.	encouraging
2004	The law covering rights for the disabled is revised.	encouraging
2004	Discrimination against people with HIV/AIDS is banned.	encouraging
2005	Government publishes first White Paper on democracy, but little is proposed in the way of political reforms.	encouraging
2006	Confucianism, previously banned, encouraged by government.	encouraging
2007	Harmonious Socialist Society proposed by government to address inequalities, but social stability is stressed over the expansion of citizen's rights.	encouraging
2008	The Beijing Olympics focused the world's attention on China's abuses of human rights.	encouraging

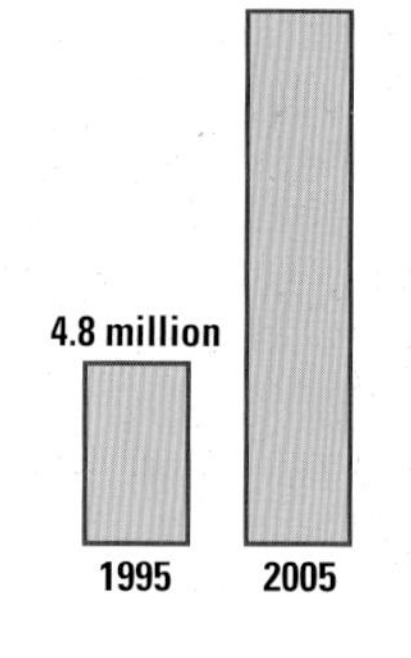

PETITIONERS

Number of people journeying to Beijing and provincial capitals to express their complaints

1995 & 2005

Petitioning was severely curtailed in the run-up to 2008 Beijing Olympics.

 Copyright © Myriad Editions

ORGANIZATIONS
And their relationship to the Party-State

PARTY-STATE

Pillar organization

The Communist Youth League of China (CYL)

The CYL is a mass youth organization under the leadership of the Chinese Communist Party (CCP). Almost all high-school graduates becoming members, and branches of the CYL have been established even in privately funded colleges and universities. The CYL is an important training ground for Party leadership.

MEMBERSHIP
2006

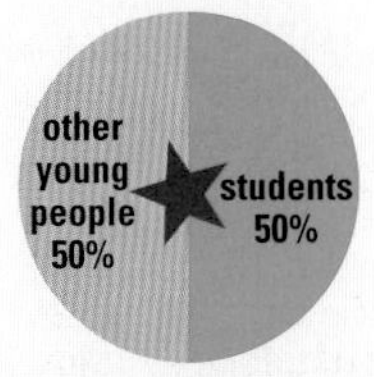

Pillar organization

All-China Federation of Trade Unions (ACFTU)

Membership: 170 million *2006*

Only legal trade union organization. Chinese trade unions organize both "horizontally", to form a federation of trade unions representing workers in a single enterprise, and "vertically", to form a national trade union representing a specific skill group.

GENDER RATIO
among membership
2006

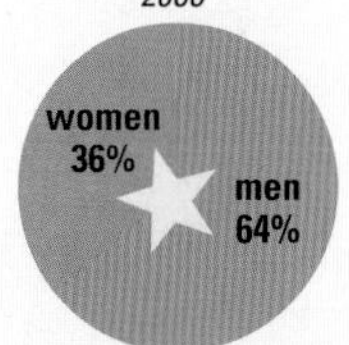

Hundreds of big corporations, such as Wal-Mart, McDonald's, Yum Brands, which operate KFC and Pizza Hut, have agreed to set up unions.

Pillar organization

All-China Women's Federation (ACWF)

Over 1 million local representatives' committees and federations

Its aim is to protect women's rights and interests and promote equality between men and women.

POLITICAL AFFILIATION
of full-time ACWF officials
2006

Total number of officials: 75,236

Officially recognized social organizations

professional organizations • educational establishments • social services • legal services • charities • foundations • cultural organizations • the five "official" religions • international NGO • advocacy groups

Unofficial social organizations

local environmental associations • women's associations • associations for older people • Qigong practice societies • neighborhood movements

Underground organizations

democracy movements • unofficial religious groups such as Chinese "house churches" • cults such as Falun Gong

These organizations are in a precarious position and not generally tolerated by the State.

Data source: *China Statistical Yearbook 2007* unless stated otherwise

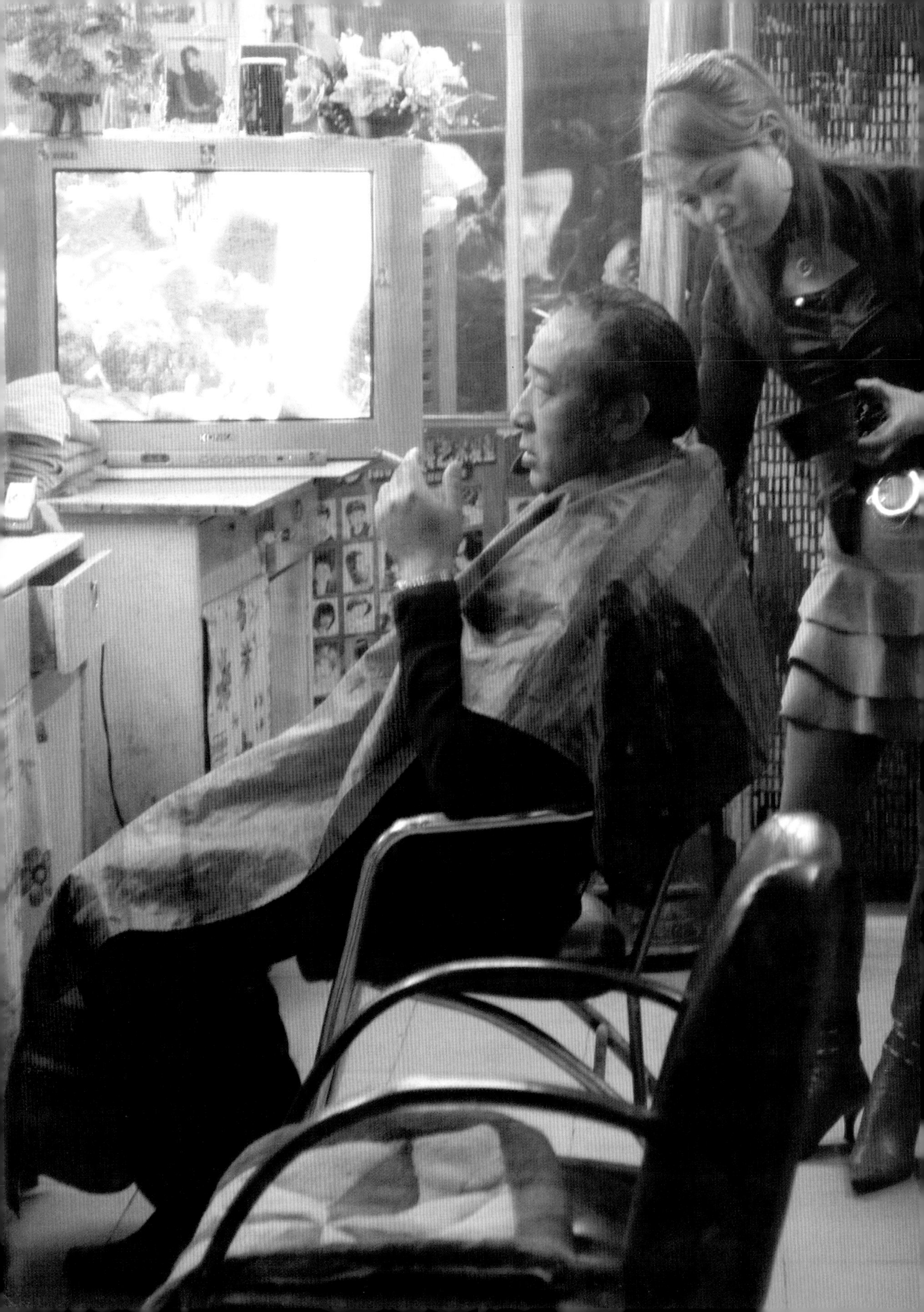

There were 30 million domestic migrants in 1982, and 150 million in 2008.

Part Six

LIVING IN CHINA

CHINA'S MODERNIZATION has entailed great changes for the people and structure of society. The rate of transformation continues apace and, unsurprisingly, there have been winners and losers. Most Chinese are richer than before, but some are a lot richer than others. Some extremely wealthy business people are now consolidating their family's politico-social status through high-profile education for their children. Other people are unemployed, or employed on subsistence wages, unsure of whether they count at all in this new entrepreneurial world. Still others, perhaps the majority in cities are struggling along as middle-income workers, facing each new financial challenge as it blows their way. For every story of progress and reform, there is likely to be a tale of suffering and regression.

The poorest of the poor are the 150 million migrant workers, male and female, drawn in the main from the 730 million Chinese who hold rural passports (*hukou*), and who come to the cities to make money that exceeds what most of them can earn from their small-holdings. Given that the government now plans to allow peasants to sell or lease land this situation may change. Some may choose to stay in rural areas and increase their stake in rural activities; others will sell up and move to the cities. The results will undoubtedly require regulating to support these people better than they have been under the antiquated *hukou* registration policies of the 1950s. As elsewhere in the world, where the rich-poor divide is stark, a government that wishes to maintain social order and harmony must find ways of reducing such inequality and alleviating its impact.

The migrant worker situation derives from the size of the rural population, the fragility of the agricultural economy, and the diversity of the country's culture. There are huge differences between the metropolitan districts, such as Beijing, Tianjin, Chongqing, and Shanghai – but this is multiplied exponentially across provinces and regions. An ordinary life in Hong Kong SAR is vastly different from that in Shaanxi or what might be "normal" in semi-autonomous region such as Xinjiang.

China faces some clearly defined challenges. Despite the achievements of the one-child policy (now being relaxed), there are still 1.3 billion people in a now ageing population, to be kept fed and healthy. In cities, parents are discovering the danger of obesity, as their children join the ranks of over-fed, under-nourished and under-exercised fast-food junkies. A major feature of life in China today is uncertainty. Religion is growing apace, with an unofficial and disputed total of 100 million Christians now rivaling established populations of Buddhists, Daoists and Muslims.

But perhaps the most obvious aspect of China's transforming society is the growth of the media. Television is ubiquitous, with sport a major draw. Internet use is growing fast, with most netizens favoring mp3 music downloads, and net-forums as their preferred use of the medium. The future of China's media, and indeed its whole creative sector, is key to a wider prognosis for Chinese society. The media operate under different rules from those practiced in liberal media zones. This does not preclude some interesting content and some good reportage, but it does leave a door open to foreign imports on the one hand, and a lack of vital information for people on the other, which again causes uncertainty and stifles potential.

EATING IN THE EAST, SLEEPING IN THE WEST

China's one-child policy is creating a population with a high proportion of elderly dependants.

In the 1980s and 1990s, the most striking change in China's households was the top-heavy familial structure caused by the one-child policy. Now, the effects of that policy, combined with the results of a semi-privatized industrial and manufacturing economy, have led to an increasing number of elderly retirees dependent on the working adult population.

This is exacerbated by the practice of early retirement with inadequate pension provision. Many urban people in their fifties have nothing to look forward to except daily pursuits such as singing classes and dancing in the park and care of their solitary grandchild.

The rise in personal property ownership is also a costly venture, and leads those in the "working middle classes" to seek more involvement in property owners' committees, whereas once they may have sought influence through residents' committee in government housing.

see also page 119

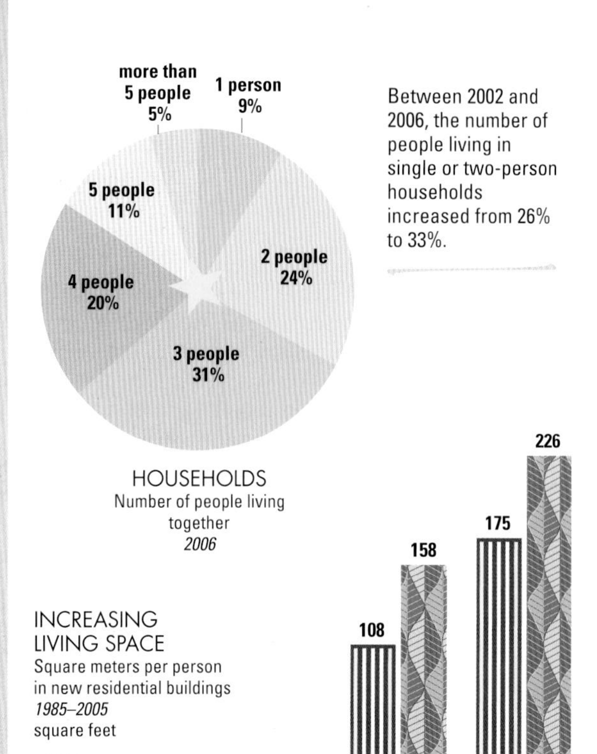

HOUSEHOLDS
Number of people living together
2006

Between 2002 and 2006, the number of people living in single or two-person households increased from 26% to 33%.

INCREASING LIVING SPACE
Square meters per person in new residential buildings
1985–2005
square feet

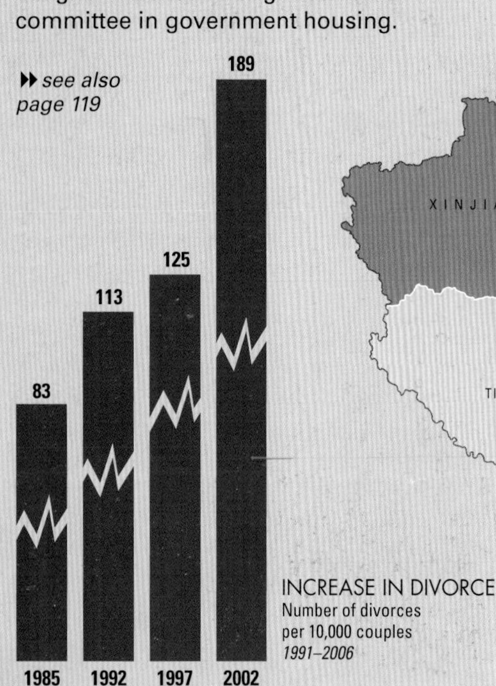

INCREASE IN DIVORCE
Number of divorces per 10,000 couples
1991–2006

DIVORCE
Number of divorces per 10,000 couples
2006

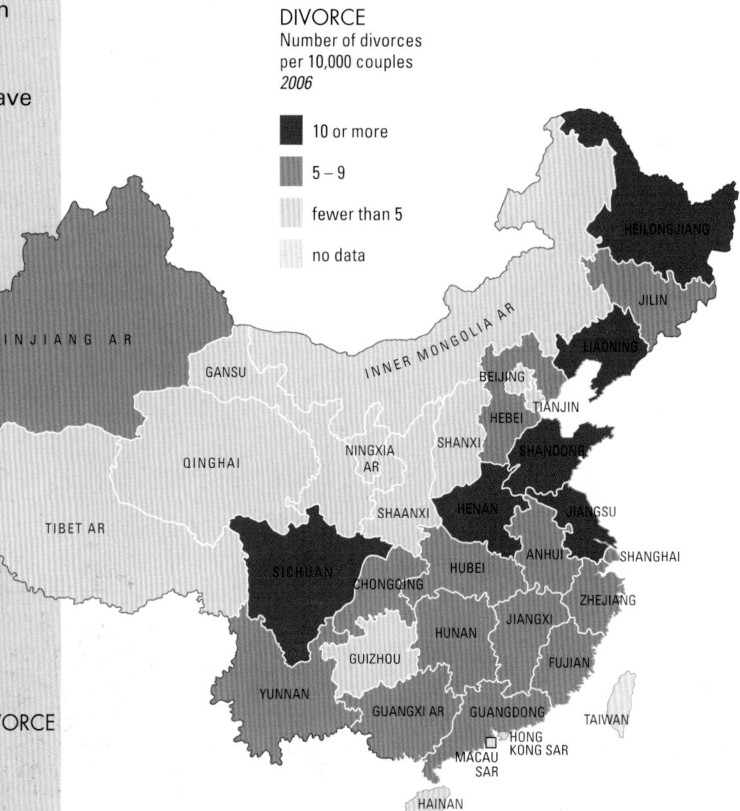

 Copyright © Myriad Editions

The falling birth rate means that by mid-century China will be facing a dependency crisis equal to that in the older industrialized countries.

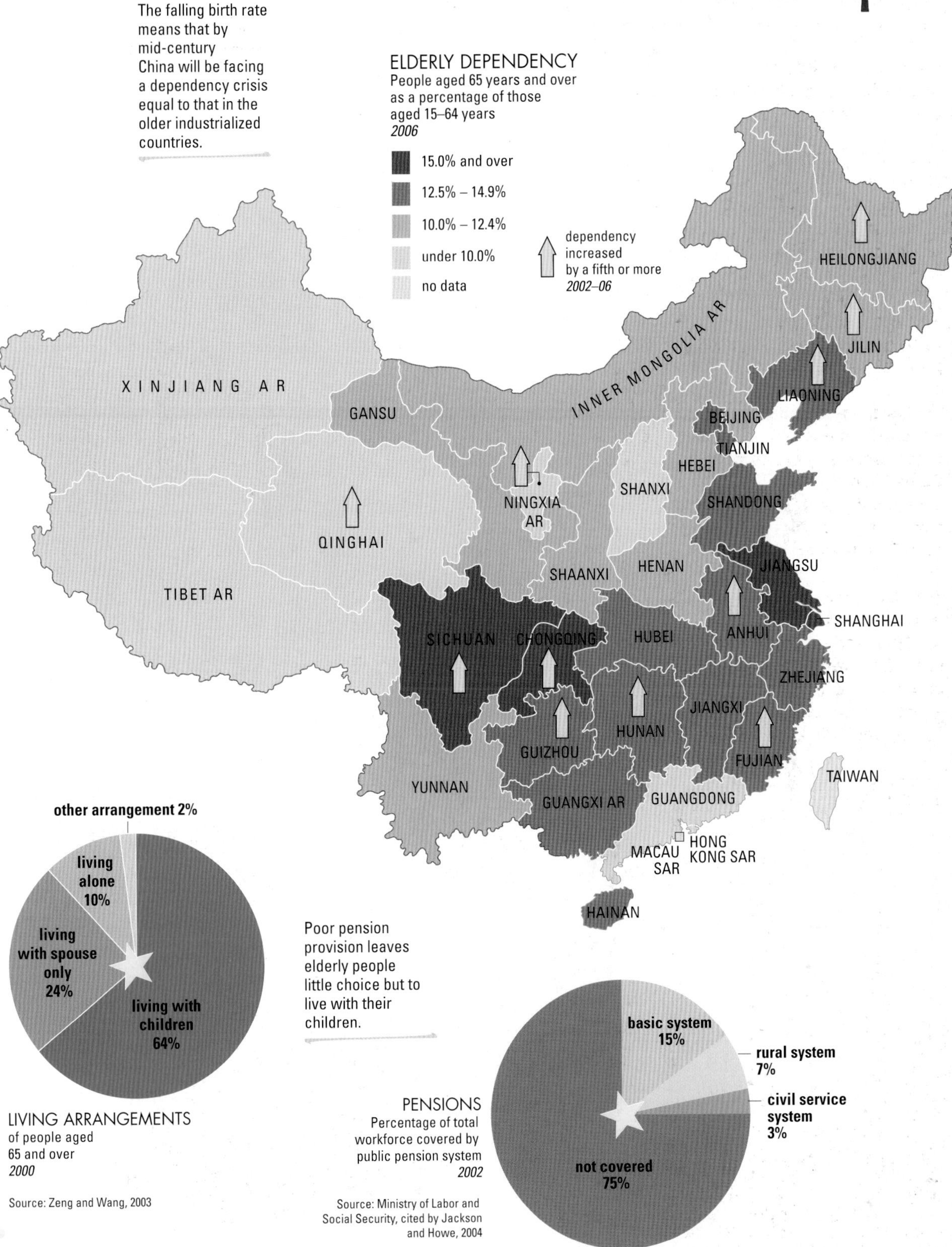

Poor pension provision leaves elderly people little choice but to live with their children.

LIVING ARRANGEMENTS
of people aged 65 and over
2000

Source: Zeng and Wang, 2003

PENSIONS
Percentage of total workforce covered by public pension system
2002

Source: Ministry of Labor and Social Security, cited by Jackson and Howe, 2004

Data source: *China Statistical Yearbook 2007* unless stated otherwise

IDEOLOGY CANNOT SUPPLY RICE

China's national diet is changing. Urbanites, in particular, are eating an increasing amount of meat and dairy products.

This is having a major impact, not only on the health of the Chinese, but on the availability and price of animal products and feed worldwide. It is one of the factors in the steep rise in food prices in 2007 and 2008.

While the children of the rural poor show signs of malnutrition, across China the average height and weight of young children has increased over a surprisingly short period. Around a quarter of urban 10–12 year olds are overweight or obese, and health and education officials are encouraging more exercise in schools and warning of the dangers of fast-food. Meanwhile, fast-food companies, such as KFC and Pizza Hut view China as a major source of profit over the next decade – although they are having to adjust their menus to suit local tastes.

Trust in Chinese food has been seriously undermined by allegations about standards of safety. The government's response has been contradictory – introducing improvements in information and enforcement, while aggressively denying that there is a problem. In 2008, however, there was no hiding the effect of the contamination of milk powder with melamine, which resulted in the death of four babies, and possible long-term consequences for the health of thousands more who fell sick.

▸▸ *see also page 120*

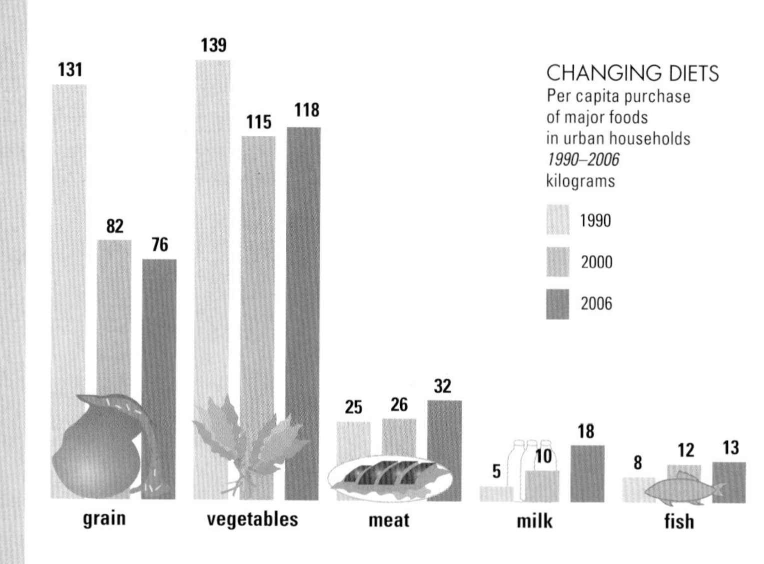

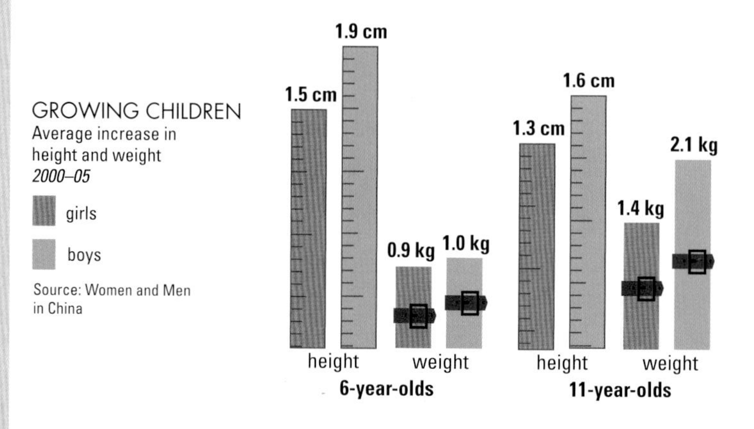

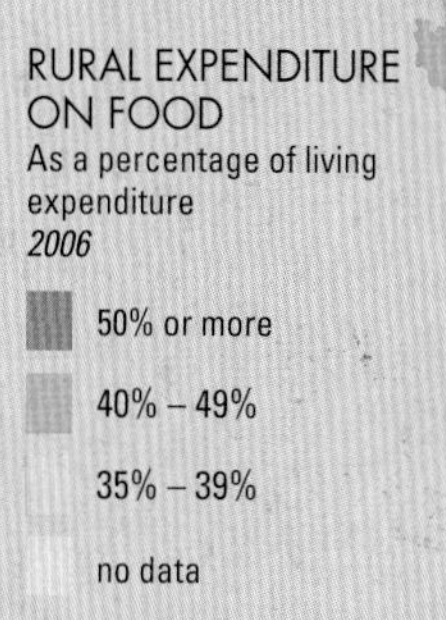

 Copyright © Myriad Editions

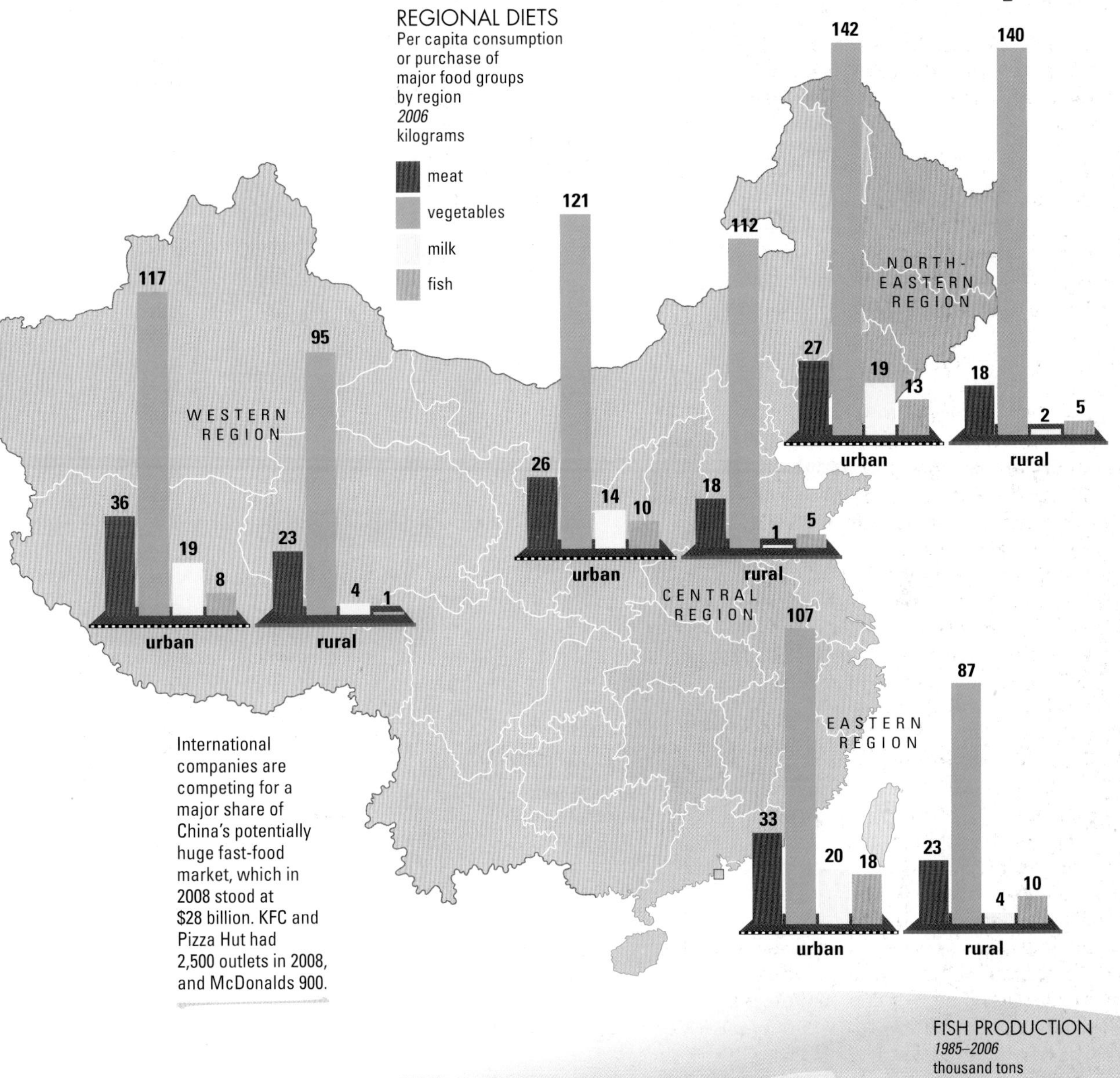

International companies are competing for a major share of China's potentially huge fast-food market, which in 2008 stood at $28 billion. KFC and Pizza Hut had 2,500 outlets in 2008, and McDonalds 900.

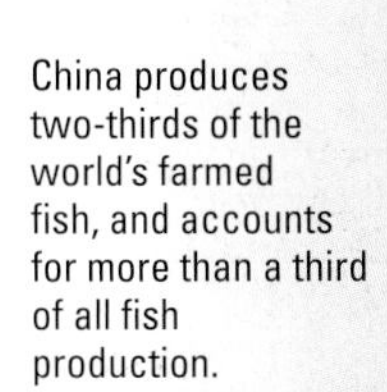

China produces two-thirds of the world's farmed fish, and accounts for more than a third of all fish production.

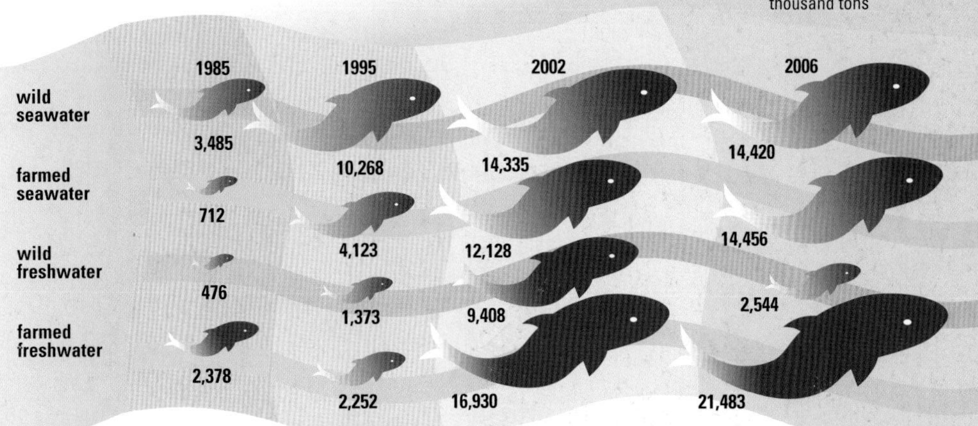

SUITING THE MEDICINE TO THE ILLNESS

China's healthcare system is managed by the state, but patients are charged fees. Increased commercialization is putting certain treatments beyond the means of many households.

Two-thirds of Chinese men smoke. While only 3 percent of women are smokers, about half of them are exposed to tobacco smoke in their homes, as are their children.

There are an estimated 700,000 people living with HIV in China, including around 75,000 AIDS patients. Although the prevalence rate is relatively low, at 0.1%, there are pockets of high infection. In 2007, unsafe sex overtook intravenous drug use as the main cause of new infection. Having been ignored by the Chinese authorities until around 2003, HIV/AIDS is now the focus of concerted government action, including targeted education, and prevention of mother-to-child transmission.

▸▸ *see also page 120*

DECLINE IN MATERNAL MORTALITY
Number of deaths per 100,000 live births
1990–2006

urban
rural

Source: Health Statistics Digest of China 2007

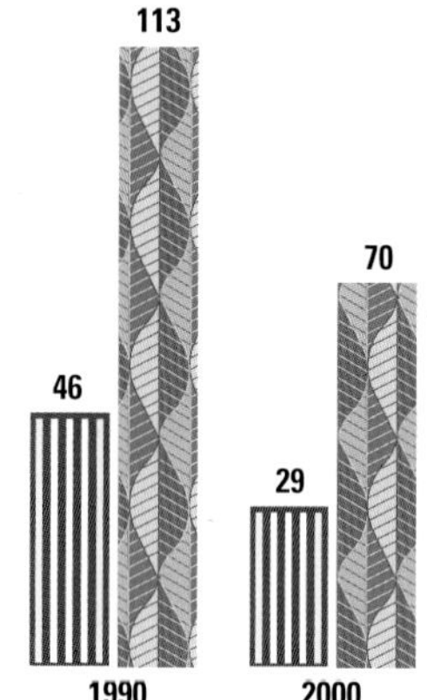

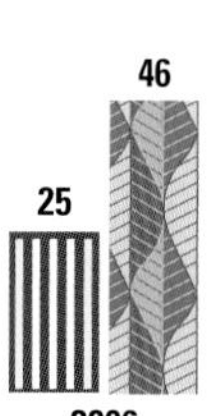

The percentage of births taking place in hospital increased from 44% to 88% between 1990 and 2006.

INCREASE IN TUBERCULOSIS
Registered cases
2000–06

women
men

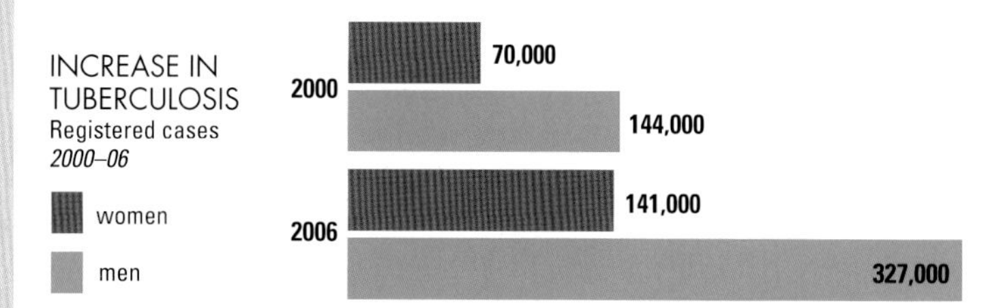

Nearly 10% of Chinese identified as HIV positive live in Xinjiang. Most are male intravenous drug users.

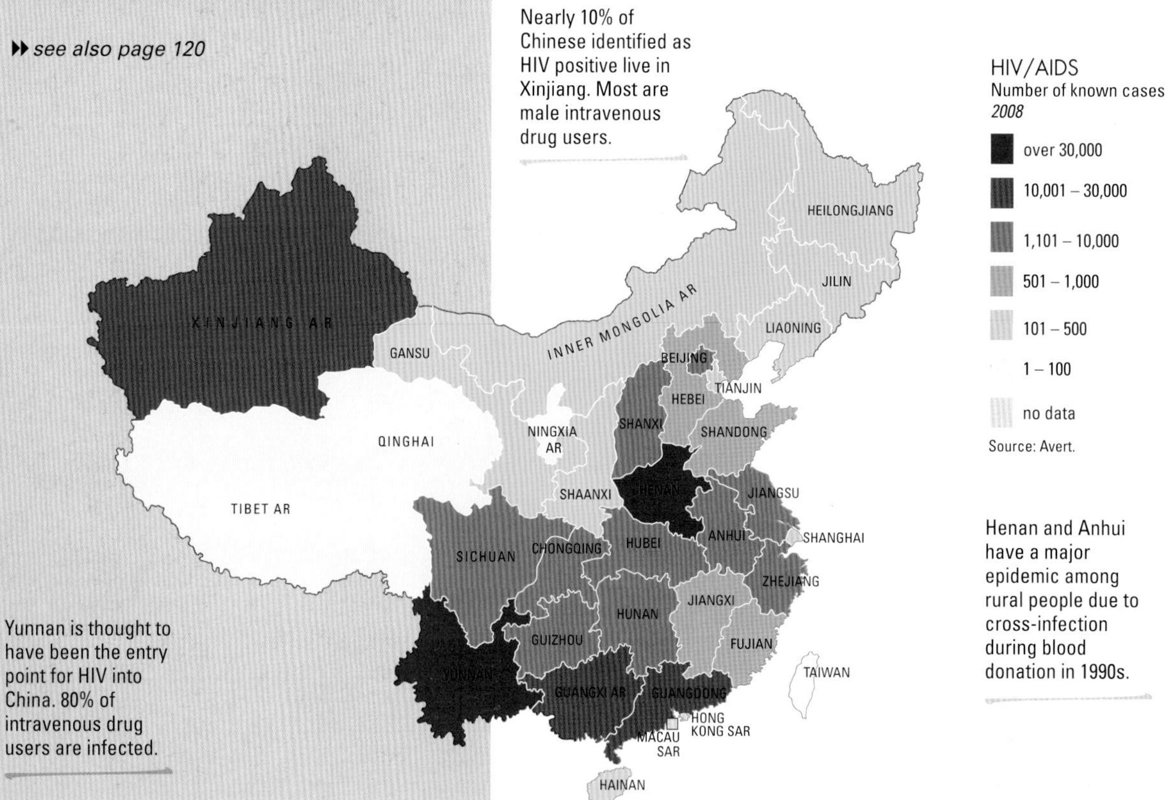

Henan and Anhui have a major epidemic among rural people due to cross-infection during blood donation in 1990s.

Yunnan is thought to have been the entry point for HIV into China. 80% of intravenous drug users are infected.

 Copyright © Myriad Editions

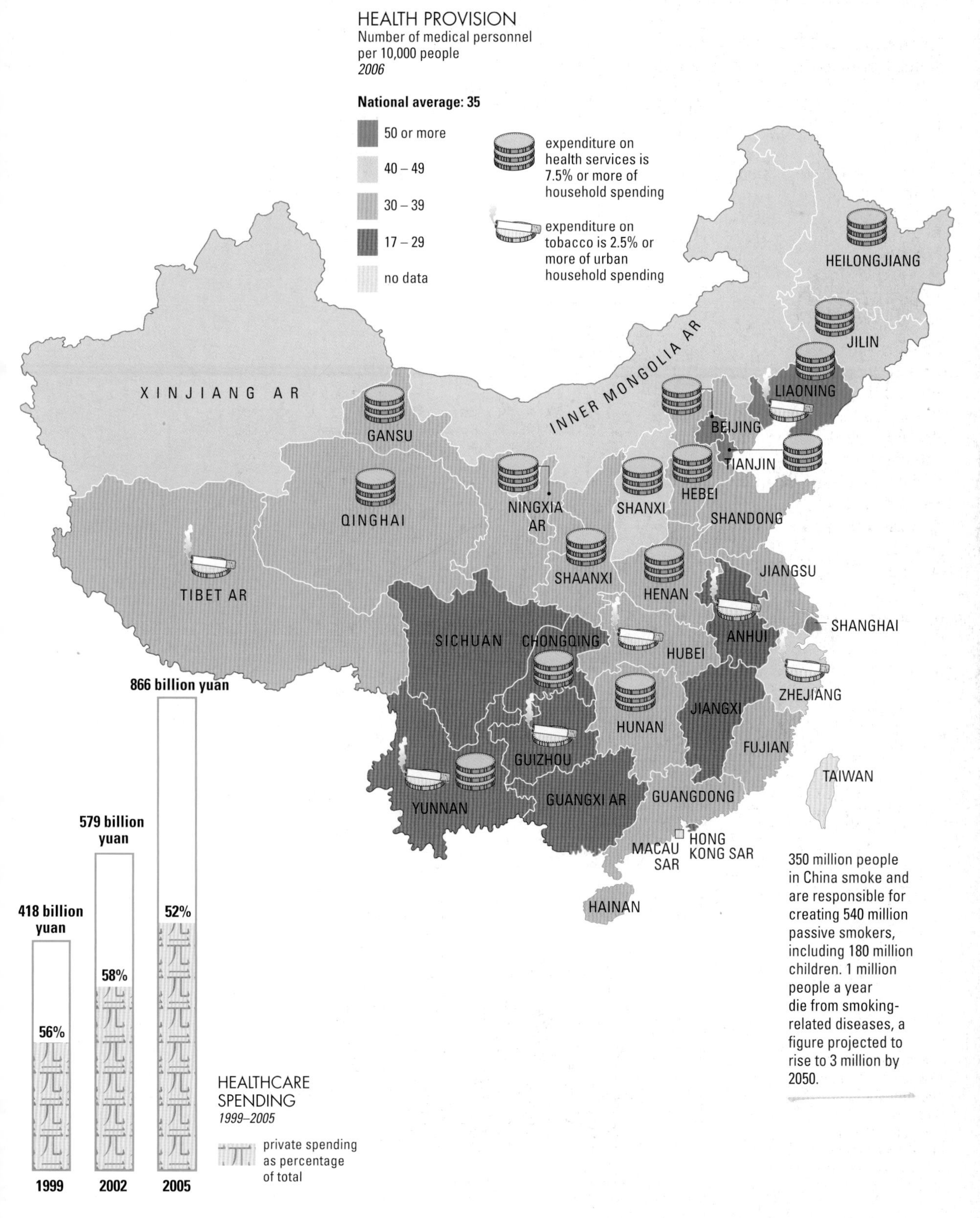

350 million people in China smoke and are responsible for creating 540 million passive smokers, including 180 million children. 1 million people a year die from smoking-related diseases, a figure projected to rise to 3 million by 2050.

Data source: *China Statistical Yearbook 2007* unless stated otherwise

GREAT VESSELS TAKE LONGER TO COMPLETE

China has over 130 million children in primary and secondary education.

Urban single children of well-off families have the best opportunities, as their parents are able to seek out the highest tier of government and private schools (*minban*). There are also "experimental schools", which service high achievers and focus on special disciplinary skills and the acquisition of foreign languages.

Competition for China's top universities is extremely fierce, but the richest parents of children who fail to get in will often pay for them to attend overseas universities, rather than one of the lower-ranking Chinese courses.

Rural and poor children have experienced very different circumstances over the course of the "Reform era", and this has recently been recognized both as a result and a cause of social inequality and dysfunction. In 2007, the government waived tuition fees for nearly 150 million rural students, and in 2008 announced plans to extend the waiver to those attending state-run schools in urban areas.

Meanwhile, many of the medalists in the 2008 Olympic Games were rural children with sports talents, whose success was underpinned by scholarships to privately sponsored sports academies.

▸▸ *see also page 121*

GOVERNMENT EXPENDITURE ON EDUCATION
1991–2006
billion yuan

Source: State Council Information Office

1991	1995	2001	2006
73	188	464	478

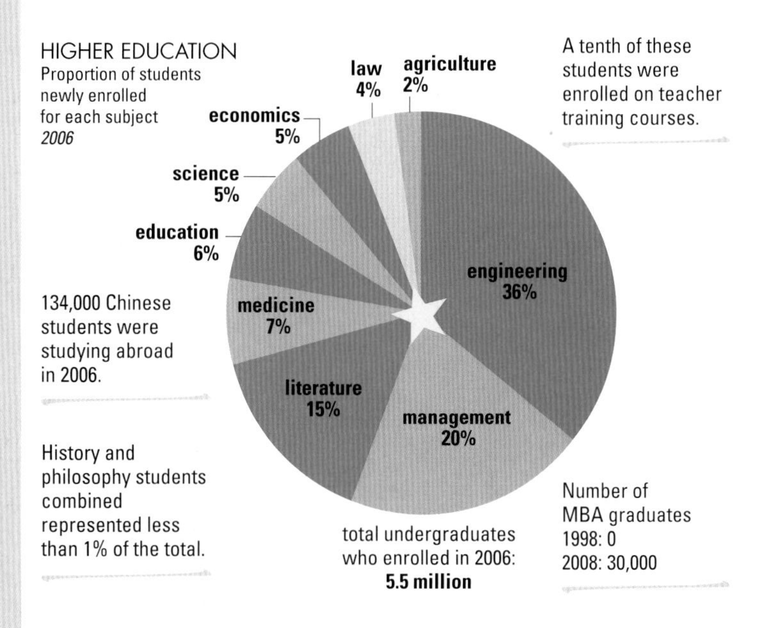

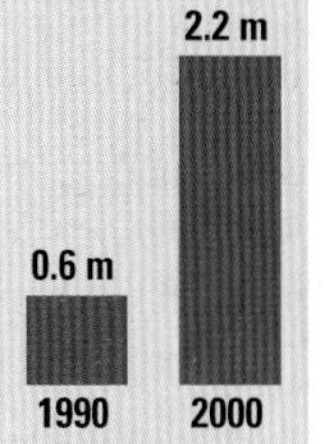

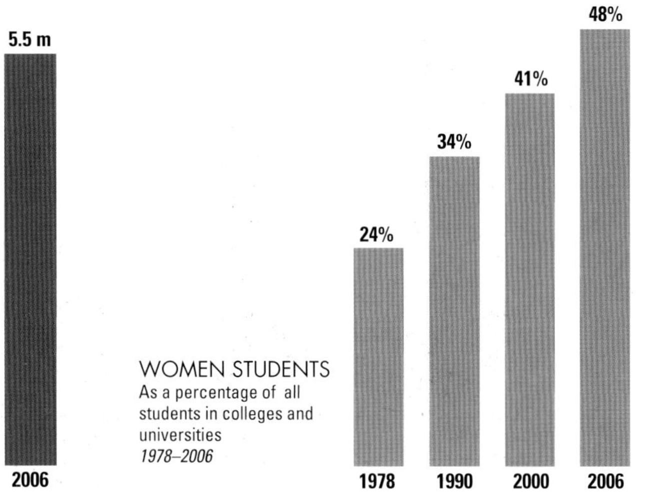

 Copyright © Myriad Editions

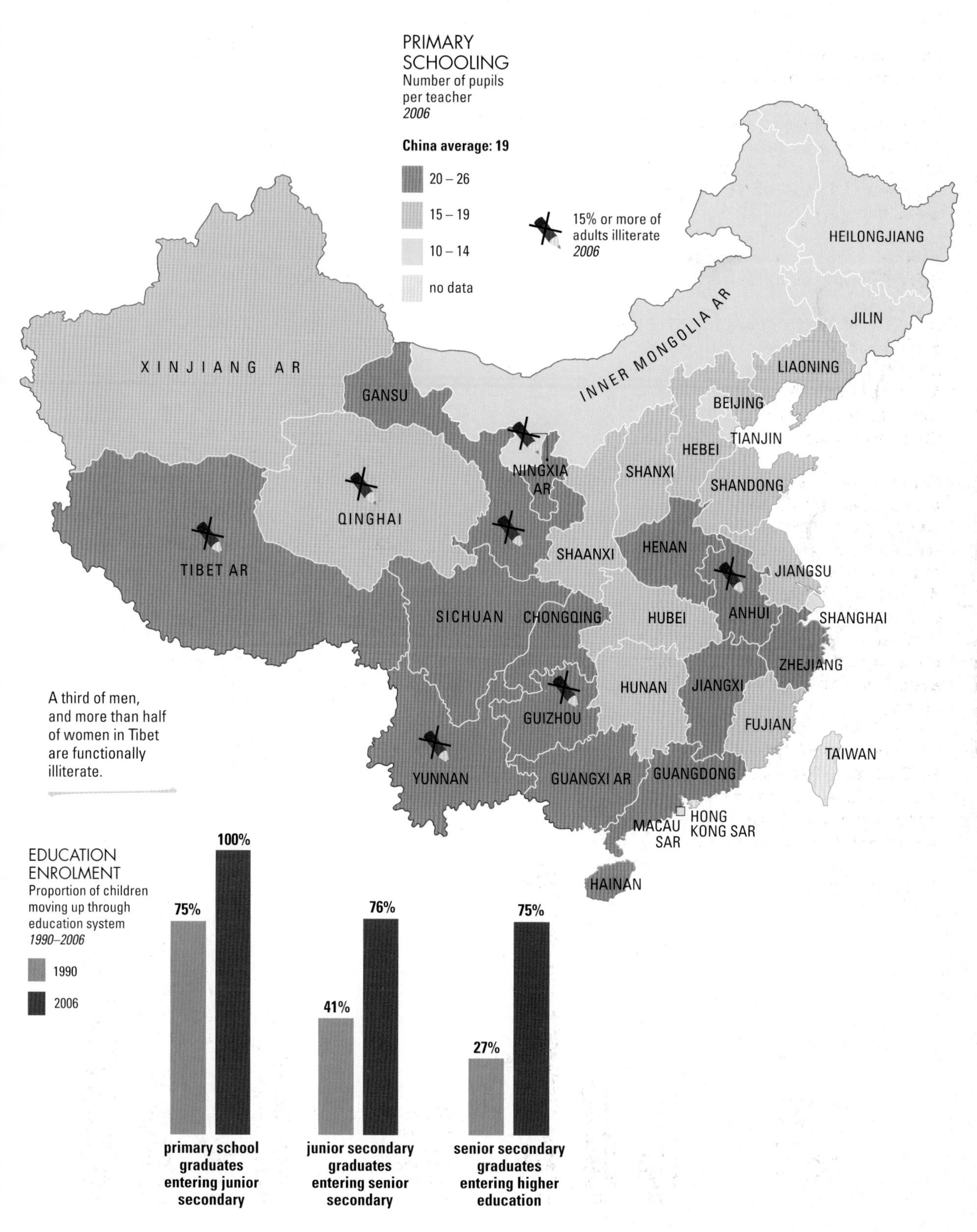

Data source: *China Statistical Yearbook 2007* unless stated otherwise

THE CROWD THAT RECEIVES

China's burgeoning media sector is a sign of its growing economy.

Advertising on television, online, in print media and on public screens and hoardings heralds the new "creative" economy. But the true creativity of China's media industries still lies in international deals for content provision, in cross-financing conglomeration, and in working through and around the prevailing routine of censorship.

Despite an increase in feature film production the industry is under threat from poor or non-existent distribution for independent film-makers, and competition for mainstream success from WTO-sanctioned imports.

The Chinese Communist Party is committed to stamping out corruption in corporate and political life, and is relying on the press to help it to do this. The dilemma is that an assertive press can also undermine Party controls.

The rise of the Chinese blogger is a phenomenon that affects both internet usage and state policy. The top blogs are important sources of information and promote discussion on all kinds of issues. It has been noted that extreme nationalism is also fostered in the blogosphere to the extent that some are closed down when the rhetoric becomes too aggressive.

▸▸ *see also page 122*

FILM ENTERTAINMENT
Total number of feature films and cartoons
1994–2006

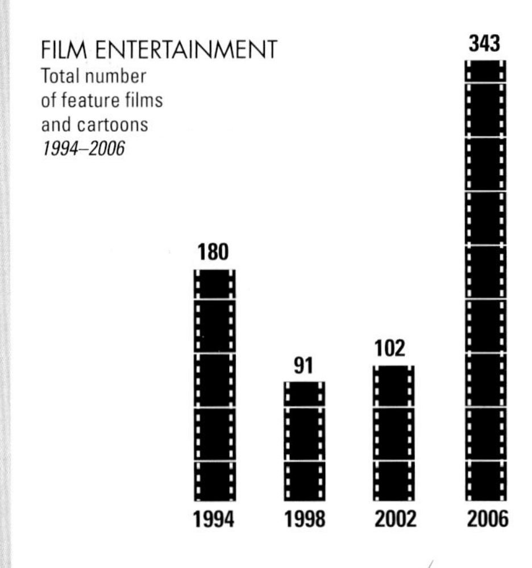

BROADCASTING
Total hours of radio and TV production
2002–06

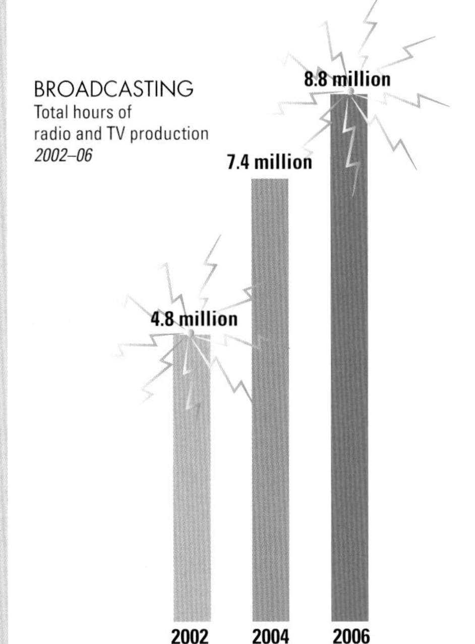

Over 160,000 new books were published in 2006, double that in 2000.

CCTV DOMINATES TV SPORT
Preferred TV channel among sports watchers
2007–mid-2008

Source: CSM Research Focus

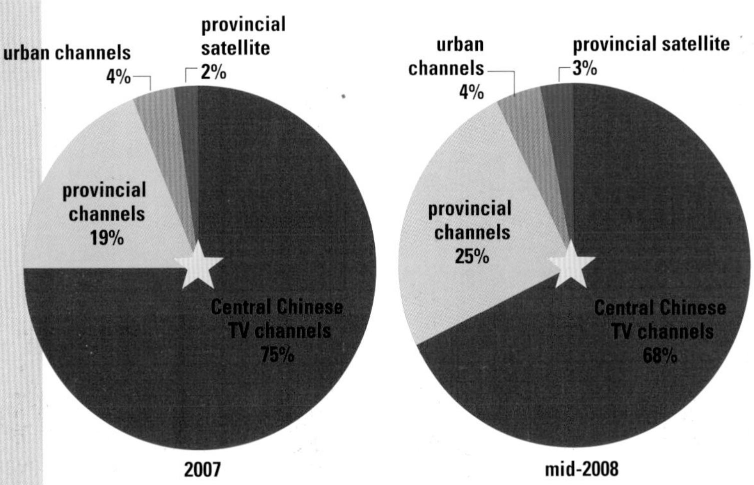

The wealthiest people watch nearly 15 minutes of sport a day, while the poorest watch less than 6 minutes.

 Copyright © Myriad Editions

Half the ten most popular Chinese blogs in October 2008 were on financial matters.

In 2008, China overtook the USA as the country with the largest online population.

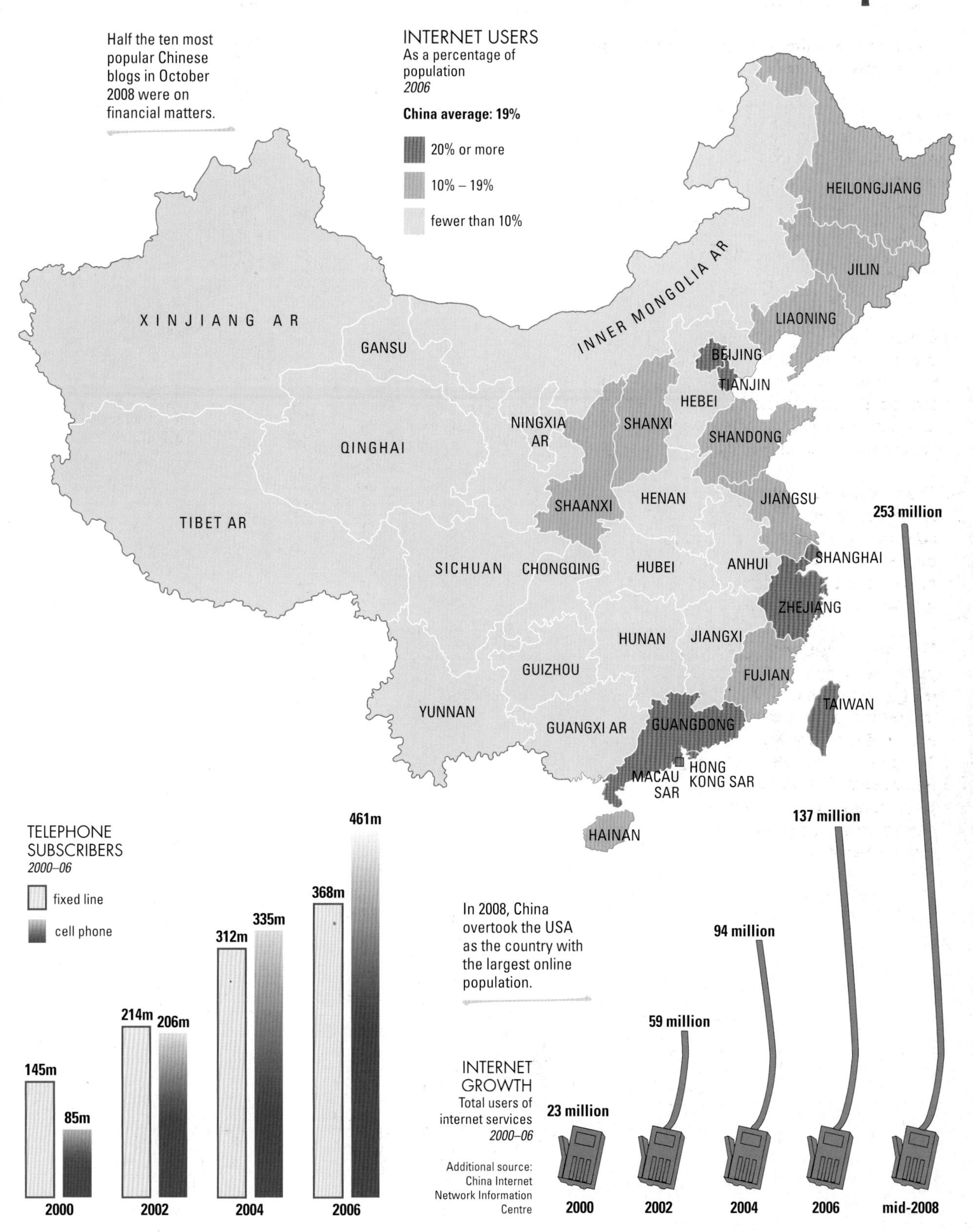

Data source: *China Statistical Yearbook 2007* unless stated otherwise

MANDATES OF HEAVEN

Only five religions are allowed by China's government.

They are tolerated under restricted conditions which prohibit the use of a religious venue or activities that might harm national or ethnic unity. Chinese indigenous religion fuses Daoism, Buddhism, folk religion and Confucianism, and is widely practised, often alongside other religions.

Whilst the ideological fervor of Maoism has largely dissipated over the past 30 years, passionate and politically meaningful affiliations remain in China today. The rise of neo-Confucianism, first observed in Singapore in the 1980s, is central to the government's dictum of harmonious society. This shift was clear in the opening ceremony of the 2008 Beijing Olympics, in which a Confucian scholar played a central role, and revolutionary ballet was notably absent.

Even more popular, perhaps, is the rise of nationalism, which gives younger people a point of reference with the older generations, and a reason for collective pride in the economic developments that have made their lifetimes so visible for the rest of the world.

▶▶ *see also page 123*

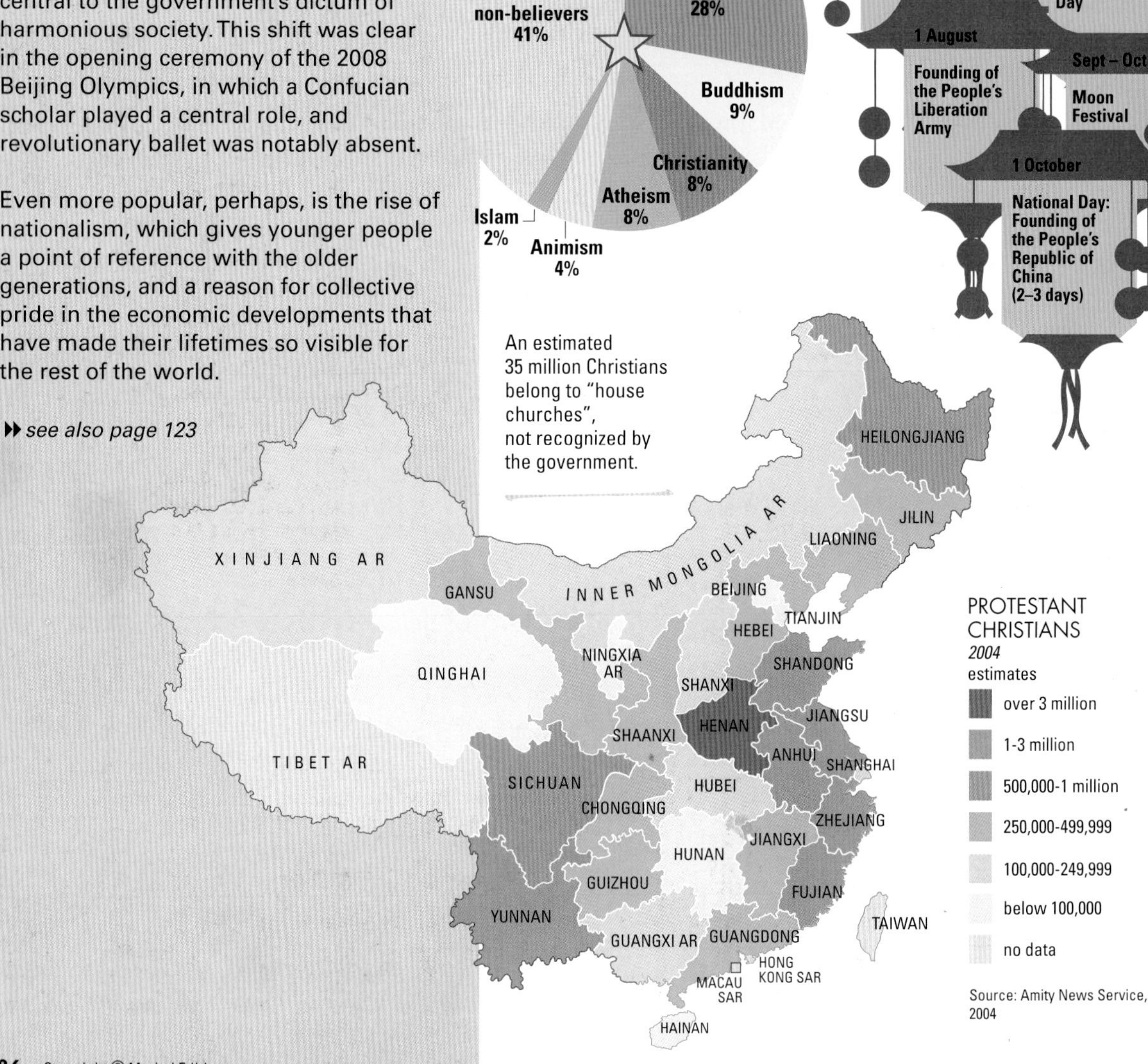

 Copyright © Myriad Editions

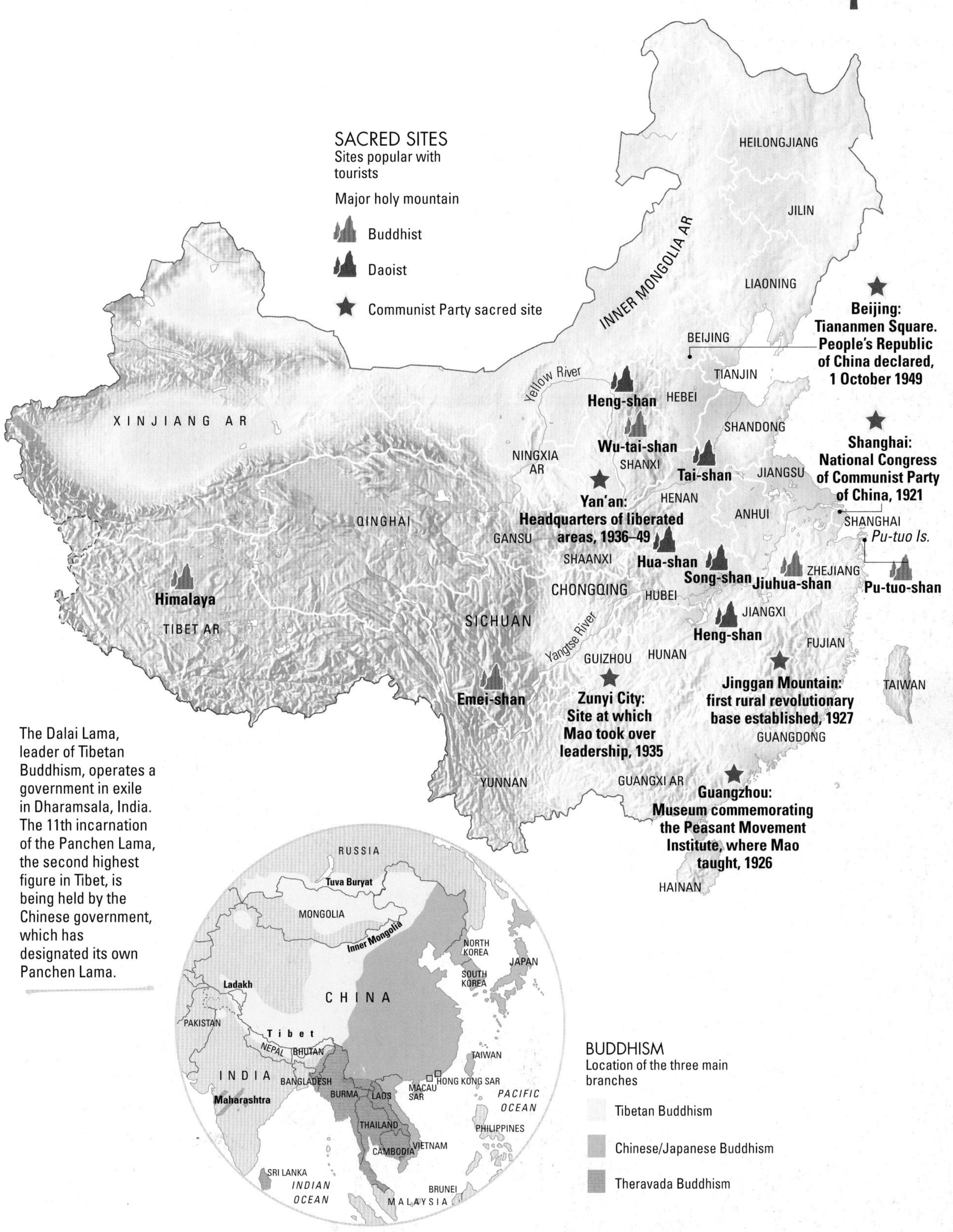

The Dalai Lama, leader of Tibetan Buddhism, operates a government in exile in Dharamsala, India. The 11th incarnation of the Panchen Lama, the second highest figure in Tibet, is being held by the Chinese government, which has designated its own Panchen Lama.

Data source: *China Statistical Yearbook 2007* unless stated otherwise

sina 新浪
aigo 爱国者

"Seek truth from facts."
Mao Zedong

Part Seven
DATES AND DATA

NO COUNTRY IN THE WORLD is more aware than China of the need for good economic and social data, which serves as a basis for policy-making, as indicators for a market-led economy, and allows commentators to assess the economic and social conditions of contemporary China. How far China has managed to gather such data, and how the data has been used, needs careful consideration.

The Great Leap Forward (1957–1961) was an unprecedented historical lesson. A series of ambitious plans, including the mass mobilization of the rural population to build infrastructure projects, and the rapid collectivization of the countryside into communes, went awry. Beijing lost control, having devolved the collection of statistics and authority for planning to the provinces and to regional levels of government. Reports of over-achievements were rife. The ensuing statistical crisis contributed to precipitating a famine in which up to 30 million people died. It was an experience not to be repeated.

The reformers, led by Deng Xiaoping, undertook a massive gamble in 1978 by dismantling much of the command economy and replacing it with a market-led economy. At the same time, it abandoned the Maoist policy of self-sufficiency and opened up China to foreign investment. But this still required planning, and planning needs reliable data that also meets the expectations of those trading and investing in China. As the modernization of China proceeds apace and the country strives to become a powerhouse driven by trade in the international community, so must its information base improve.

This book has used two main sources of official data. The *China Statistical Yearbook* contains a wealth of information about almost every aspect of China: employment, agriculture, trade, schooling, healthcare, housing. More detailed, and more intimate, data are given in the *China Population Statistics Yearbook*, including the number of rooms in a dwelling, and the main form of contraception used by householders. Much of this is extrapolated from the 2000 Census, the fifth in a 10-yearly series.

Taking a census of China's 1.3 billion people is no mean feat. The 2000 Census used 10,000 tons of paper for the questionnaires, and 5 million enumerators to collect the data. It was most notable, however, for the innovations introduced since the previous census. For the first time, the questionnaire included a confidentiality clause, presumably intended to encourage respondents to be more truthful. Unusually detailed questions were asked about housing – the age and condition of the property, its water supply and sanitary facilities, the length of tenure and the weekly rent – information clearly of interest to a government working through the transition from a state-controlled to a market-led housing policy. The issue of unemployment and welfare support was addressed openly for the first time. And the timing of the census was shifted from 1 July to 1 November, with the result that migrant workers were more likely to have been enumerated at their place of work, rather than in their home town or village.

Officially produced data has two uses: as a basis for policy-making and as a way of presenting China in the best light possible. The information the Party-State chooses to collect is likely to support its promotion of a market-led economy.

We have relied mainly on official data, well aware of their strengths and weaknesses. We are confident, however, of their usefulness in providing an overview, and as a guide to trends and developments in one of the most diverse and rapidly changing countries in the world.

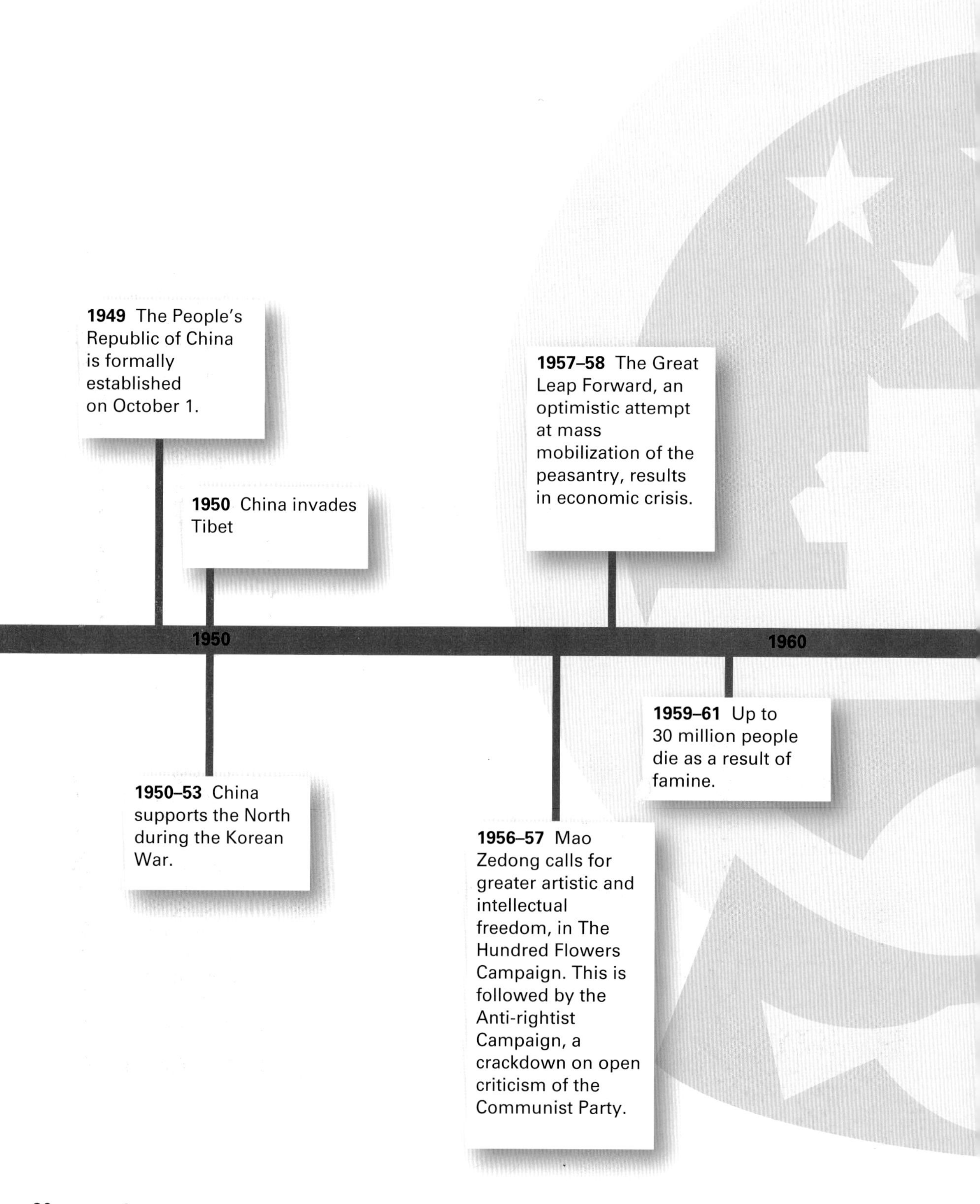

 Copyright © Myriad Editions

1966–76 The Cultural Revolution takes place, inspired by Mao Zedong's campaign to advance socialism and curb what he views as capitalist tendencies within the Communist Party. Hundreds of Tibetan monasteries are destroyed.

1970

1971 China is admitted to the United Nations.

1972 US President Nixon visits China.

1976 Mao Zedong and Zhou Enlai die.

1978 Deng Xiaoping is made "paramount leader" and announces economic reforms. Writings posted on what comes to be known as the "Democracy Wall" in Beijing call for political reforms.

1979 USA–China relations are normalized. The small-family policy is announced.

1980

1980 China joins the World Bank and the International Monetary Fund.

1982 USA sells weapons to Taiwan, but the USA and China make agreement on fewer future sales.

1987 The Dalai Lama is awarded the Nobel Peace Prize for leading non-violent struggle for Tibet.

1989 More than 100,000 people protest in Tiananmen Square, and are brutally suppressed. Jiang Zemin becomes General Secretary of the Communist Party.

1992 Deng Xiaoping tours southern provinces, pressing economic reforms.

1994 USA extends China's most-favored nation status, which includes expectations of improvement in China's human rights performance, but no preconditions.

1980

1990

 Copyright © Myriad Editions

CHRONOLOGY 1980–2010 中

1997 Deng Xiaoping dies. Hong Kong reverts to China.

1999 Macau reverts to China.

2000 Jiang Zemin's "Three Represents" opens the Communist Party membership to entrepreneurs, intellectuals and scientists.

2001 China joins the World Trade Organization.

2002–03 In a smooth transition, Hu Jintao becomes General Secretary of the Communist Party and President of the People's Republic of China.

2003 The Three Gorges Dam begins operation.

2006 China–Africa Summit in Beijing is attended by heads of 48 African nations. China promises billions of dollars in loans and credit.

2006 A resolution on building a "Harmonious Socialist Society" addresses the inequalities created by economic growth. First railroad to Tibet is opened.

2007 Amnesty International claims China has failed to keep its promise to improve human rights in the run-up to the Olympic Games.

2008 Sichuan earthquake leaves up to 80,000 dead or missing. The Olympics take place in Beijing.

2010 The World Exposition is to take place in Shanghai.

2000

2010

Selected countries	1 Population			2 Life expectancy at birth	3 Gross Domestic Product		
	Total *2006* million	Annual growth rate *2006*	Density *2006* people per km²	*2006*	Total *2007* US$ billion	Annual growth *2007*	Per person International $ *2007*
Argentina	39	1.0%	14	75.03	262	8.7%	13,244
Australia	21	1.5%	3	81	822	4.5%	34,882
Bangladesh	156	1.8%	1,198	63.66	68	6.5%	1,242
Brazil	189	1.3%	22	72.08	1,314	5.4%	9,570
Canada	33	1.0%	4	80.36	1,326	2.7%	35,729
China	**1,312**	**0.6%**	**141**	**72**	**3,280**	**11.9%**	**5,345**
Czech Republic	10	0.4%	133	76.48	168	5.6%	23,194
Egypt	74	1.8%	75	71.01	128	7.1%	5,352
France	61	0.6%	111	80.56	2,562	2.2%	33,414
Germany	82	–0.1%	236	79.13	3,297	2.5%	33,154
India	1,110	1.4%	373	64.47	1,171	9.0%	2,753
Indonesia	223	1.1%	123	68.16	433	6.3%	3,728
Iran	70	1.5%	43	70.65	271	7.6%	10,934
Israel	7	1.8%	326	80.02	162	5.3%	25,917
Italy	59	0.4%	200	81.08	2,107	1.5%	29,935
Japan	128	0.0%	350	82.32	4,377	2.1%	33,525
Kazakhstan	15	1.1%	6	66.16	104	8.5%	10,829
Malaysia	26	1.8%	79	74.05	181	5.7%	13,380
Mexico	104	1.1%	54	74.47	893	3.3%	12,780
Mongolia	3	1.2%	2	67.17	4	9.9%	3,222
New Zealand	4	1.2%	16	79.93	129	3.4%	26,108
Nigeria	145	2.4%	159	46.78	166	6.3%	1,977
Pakistan	159	2.1%	206	65.21	144	6.4%	2,525
Philippines	86	2.0%	289	71.39	144	7.3%	3,410
Poland	38	–0.1%	124	75.14	420	6.5%	15,811
Romania	22	–0.2%	94	72.18	166	6.2%	11,394
Russia	143	–0.5%	9	65.56	1,291	8.1%	14,743
Singapore	4	3.2%	6,508	79.85	161	7.7%	50,304
South Africa	47	1.1%	39	50.71	278	4.8%	9,736
South Korea	48	0.3%	490	78.5	970	5.0%	24,712
Thailand	63	0.7%	124	70.24	246	4.8%	8,138
Turkey	73	1.3%	95	71.49	657	4.5%	12,481
United Kingdom	61	0.5%	250	79.14	2,728	3.0%	33,535
USA	299	1.0%	33	77.85	13,811	2.2%	45,790
Vietnam	84	1.2%	271	70.85	71	8.5%	2,600
World	**6,538**	**1.2%**	**50**	**68.24**	**54,347**	**3.8%**	**9,896**

 Sources: **Col 1** *China Statistical Yearbook 2008*; **Cols 2, 3** World Development Indicators; **Col 4** *CSY 2008*; Col 5 WDI; **Cols 6, 7** *CSY 2008*

4 Employment share by sector			5 World trade		6 Trade with China		7 Foreign Direct Investment in China	Selected countries
Agriculture *2005* or latest percent	Industry *2005* or latest percent	Services *2005* *or latest* percent	Imports *2005* US$ billion	Exports *2005* US$ billion	China's exports *2007* US$ million	China's imports *2007* US$ million	*2007* US$ million utilized	
1%	24%	75%	35	46	3,566	6,335	11	Argentina
4%	21%	75%	143	126	17,990	25,840	354	Australia
52%	14%	35%	14	10	3,345	114	1	Bangladesh
21%	21%	58%	102	134	11,372	18,342	32	Brazil
9%	34%	57%	386	429	19,356	10,979	397	Canada
3%	**22%**	**75%**	**712**	**837**	**-**	**-**	**-**	**China**
45%	24%	31%	86	90	4,135	831	15	Czech Republic
4%	40%	57%	29	27	4,433	240	7	Egypt
30%	20%	50%	576	556	20,327	13,341	456	France
2%	30%	68%	990	1,135	48,714	45,383	734	Germany
44%	18%	38%	184	161	24,011	14,617	34	India
25%	30%	45%	86	97	12,601	12,395	134	Indonesia
2%	22%	76%	57	74	7,284	13,306	7	Iran
4%	31%	65%	58	58	3,656	1,654	25	Israel
4%	28%	66%	462	461	21,170	10,211	348	Italy
32%	18%	50%	589	652	102,009	133,942	3,589	Japan
15%	30%	55%	26	31	7,446	6,432	2	Kazakhstan
15%	26%	59%	137	161	56,099	103,752	397	Malaysia
40%	17%	43%	242	230	17,689	28,697	6	Mexico
3%	20%	73%	2	1	11,706	3,263	1	Mongolia
7%	22%	71%	33	30	683	1,352	64	New Zealand
43%	20%	37%	35	52	2,160	1,538	12	Nigeria
37%	15%	48%	21	17	3,796	537	2	Pakistan
17%	29%	53%	51	47	5,789	1,104	195	Philippines
32%	30%	38%	114	113	7,498	23,118	9	Poland
10%	30%	60%	43	33	6,553	1,112	30	Romania
–	30%	70%	164	269	2,084	281	52	Russia
8%	27%	65%	251	285	28,466	19,689	3,185	Singapore
10%	25%	65%	69	67	29,620	17,524	69	South Africa
5%	30%	65%	316	335	7,428	6,618	3,678	South Korea
34%	23%	37%	132	130	11,973	22,665	89	Thailand
43%	20%	37%	123	106	10,476	1,292	10	Turkey
30%	25%	46%	674	593	31,656	7,776	831	United Kingdom
19%	24%	56%	2,020	1,303	232,677	69,391	2,616	USA
1%	22%	76%	39	37	11,891	3,226	1	Vietnam
-	**-**	**-**	**12,891**	**12,957**	**1,217,776**	**955,950**	**-**	**World**

Provinces	**1 Population** *2007* million	**2 Annual population growth rate** *2007*	**3 Number of girls born per 100 boys** *2000*	**4 Life expectancy** *2000*
China	**1,321.3**	**0.52%**	**86**	**71**
Anhui	61.2	1.18%	78	72
Beijing	16.3	1.13%	90	76
Chongqing	28.2	0.98%	87	72
Fujian	35.8	0.89%	85	73
Gansu	26.2	0.88%	87	68
Guangdong	94.5	0.82%	77	73
Guangxi AR	47.7	0.79%	80	71
Guizhou	37.6	0.73%	93	66
Hainan	8.5	0.69%	74	73
Hebei	69.4	0.67%	88	73
Heilongjiang	38.2	0.66%	91	72
Henan	93.6	0.65%	84	72
Hubei	57.0	0.64%	78	71
Hunan	63.6	0.60%	79	71
Inner Mongolia AR	24.1	0.53%	92	70
Jiangsu	76.3	0.53%	86	74
Jiangxi	43.7	0.50%	87	69
Jilin	27.3	0.49%	90	73
Liaoning	43.0	0.48%	89	73
Ningxia AR	6.1	0.45%	92	70
Qinghai	5.5	0.41%	91	66
Shaanxi	37.5	0.38%	82	70
Shandong	93.7	0.34%	89	74
Shanghai	18.6	0.32%	90	78
Shanxi	33.9	0.30%	89	72
Sichuan	81.3	0.29%	86	71
Tianjin	11.2	0.25%	89	75
Tibet AR	2.8	0.25%	97	64
Xinjiang AR	21.0	0.23%	94	67
Yunnan	45.1	0.21%	92	66
Zhejiang	50.6	0.15%	88	75
Hong Kong SAR	6.9	0.27%	–	82
Macao SAR	0.5	0.57%	–	82

Sources: **Cols 1, 2** *China Statistical Yearbook 2008;* **Cols 3, 4** *China Population Statistics Yearbook 2002*

5 Dependency People aged 65 and over as % of those aged 15–64 *2006*	6 Urbanization Urban population as % of population *2007*	7 Ethnicity Minority nationalities as % of total population *2000*	8 Literacy % of people aged 15 and over who are illiterate or semi-literate *2006*	Provinces
12.7%	**45%**	**8%**	**9%**	**China**
14.9%	39%	1%	16%	Anhui
14.3%	85%	4%	4%	Beijing
16.6%	48%	7%	10%	Chongqing
12.9%	49%	2%	11%	Fujian
10.4%	32%	9%	22%	Gansu
9.8%	63%	2%	5%	Guangdong
13.1%	36%	38%	6%	Guangxi AR
12.5%	28%	38%	19%	Guizhou
12.5%	47%	17%	10%	Hainan
11.3%	40%	4%	6%	Hebei
10.4%	54%	5%	5%	Heilongjiang
11.4%	34%	1%	9%	Henan
13.4%	44%	4%	10%	Hubei
14.8%	40%	10%	7%	Hunan
10.3%	50%	21%	9%	Inner Mongolia AR
15.0%	53%	0%	9%	Jiangsu
12.7%	40%	0%	9%	Jiangxi
10.5%	53%	9%	5%	Jilin
13.8%	59%	16%	4%	Liaoning
8.5%	44%	35%	15%	Ningxia AR
9.9%	40%	46%	19%	Qinghai
12.3%	41%	1%	9%	Shaanxi
12.7%	47%	1%	9%	Shandong
18.6%	89%	1%	5%	Shanghai
9.6%	44%	0%	4%	Shanxi
16.4%	36%	5%	13%	Sichuan
13.6%	76%	3%	4%	Tianjin
9.3%	28%	94%	46%	Tibet AR
9.4%	39%	59%	7%	Xinjiang AR
10.8%	32%	33%	17%	Yunnan
13.4%	57%	1%	10%	Zhejiang
–	–	–	–	Hong Kong SAR
–	–	–	–	Macao SAR

Sources: **Col 5, 7, 8** *China Statistical Yearbook 2007;* **Col 6** *China Statistical Yearbook 2008*

Provinces	9 **Gross Domestic Product per person** *2007* yuan	10 **Value of commodity imports and exports** *2006* US$ million	11 **Foreign Direct Investment** utilized *2007* US$ billion	12 **Foreign exchange earnings from tourism** *2006* US$ million
China	**18,934**	**1,760,396**	**2,109**	**33,949**
Anhui	12,037	12,245	24	227
Beijing	57,277	158,037	88	4,026
Chongqing	14,640	5,470	20	309
Fuijan	25,828	62,660	103	1,471
Gansu	10,326	3,825	3	63
Guangdong	32,897	527,199	351	7,533
Guangxi AR	12,491	6,668	22	423
Guizhou	7,288	1,618	3	115
Hainan	14,477	2,846	94	229
Hebei	19,746	18,531	29	243
Heilongjiang	18,475	12,857	14	492
Henan	16,039	9,795	26	274
Hubei	16,197	11,762	31	320
Hunan	14,477	7,352	24	503
Inner Mongolia AR	25,327	5,961	17	404
Jiangsu	33,759	283,978	382	2,787
Jiangxi	12,592	6,195	29	140
Jilin	19,358	7,914	31	137
Liaoning	25,648	48,390	109	934
Ningxia AR	14,577	1,437	2	2
Qinghai	14,196	652	2	13
Shaanxi	14,583	5,360	16	511
Shandong	27,721	95,214	96	1,014
Shanghai	65,602	227,524	257	3,904
Shanxi	16,898	6,627	18	164
Sichuan	12,926	11,021	27	395
Tianjin	45,295	64,462	83	626
Tibet AR	12,049	328	1	61
Xinjiang AR	16,817	9,103	3	128
Yunnan	10,504	6,225	12	658
Zhejiang	37,115	139,142	146	2,133
Hong Kong SAR	205,853*	–	45*	–
Macao SAR	248,645**	–	–	–
Taiwan	117,986**	–	–	–

 Sources: **Cols 9, 11** *China Statistical Yearbook 2008;* **Cols 10, 12** *China Statistical Yearbook 2007*

*converted from HK dollars; ** converted from US dollars

13 Agriculture		14 Industry		15 Services		Provinces
Value *2006* billion yuan	Output as % of GDP *2006*	Value *2006* billion yuan	Output as % of GDP *2006*	Value *2006* billion yuan	Output as % of GDP *2006*	
2,474	**12%**	**10,316**	**49%**	**8,297**	**39%**	**China**
103	17%	265	43%	247	40%	Anhui
10	1%	219	28%	558	71%	Beijing
43	12%	150	43%	156	45%	Chongqing
90	12%	374	49%	297	39%	Fuijan
33	15%	104	46%	90	40%	Gansu
158	6%	1,343	51%	1,120	43%	Guangdong
103	21%	188	39%	192	40%	Guangxi AR
39	17%	98	43%	91	40%	Guizhou
34	33%	29	27%	42	40%	Hainan
161	14%	612	52%	394	34%	Hebei
74	12%	337	54%	209	34%	Heilongjiang
205	16%	672	54%	372	30%	Henan
114	15%	337	44%	308	41%	Hubei
133	18%	315	42%	308	41%	Hunan
65	14%	233	49%	181	38%	Inner Mongolia AR
155	7%	1,225	57%	785	36%	Jiangsu
79	17%	232	50%	156	33%	Jiangxi
67	16%	192	45%	169	39%	Jilin
98	11%	473	51%	355	38%	Liaoning
8	11%	35	49%	28	40%	Ningxia AR
7	11%	33	52%	24	38%	Qinghai
49	11%	244	54%	159	35%	Shaanxi
214	10%	1,275	58%	719	33%	Shandong
9	1%	503	49%	524	51%	Shanghai
28	6%	275	58%	173	36%	Shanxi
160	18%	378	44%	327	38%	Sichuan
12	3%	249	57%	175	40%	Tianjin
5	17%	8	28%	16	55%	Tibet AR
53	17%	146	48%	106	35%	Xinjiang AR
75	19%	171	43%	154	39%	Yunnan
93	6%	851	54%	631	40%	Zhejiang
–	0%		8%	–	88%	Hong Kong SAR
–	–		19%	–	85%	Macao SAR
–	1%	–	28%	–	71%	Taiwan

Sources: **Cols 13, 14, 15** *China Statistical Yearbook 2007*

Provinces	16 Rural household consumption as % of urban 2006	17 Expenditure on food as % of living expenditure		18 Employment	
		Rural 2006	Urban 2006	Average wage 2006 yuan	Urban unemployment rate 2006
China	**-**	**43%**	**36%**	**24,932**	**4.1%**
Anhui	31%	43%	42%	17,949	4.2%
Beijing	41%	33%	31%	40,117	2.0%
Chongqing	26%	52%	36%	19,215	4.0%
Fujian	37%	45%	39%	19,318	3.9%
Gansu	23%	47%	35%	17,246	3.6%
Guangdong	28%	49%	36%	26,186	2.6%
Guangxi AR	31%	50%	42%	18,064	4.1%
Guizhou	19%	52%	39%	16,815	4.1%
Hainan	36%	53%	43%	15,890	3.6%
Hebei	30%	37%	34%	16,590	3.8%
Heilongjiang	34%	35%	33%	16,505	4.3%
Henan	30%	41%	33%	16,981	3.5%
Hubei	31%	47%	39%	16,048	4.2%
Hunan	32%	49%	35%	17,850	4.3%
Inner Mongolia AR	31%	39%	30%	18,469	4.1%
Jiangsu	43%	42%	36%	23,782	3.4%
Jiangxi	35%	49%	40%	15,590	3.6%
Jilin	36%	40%	33%	16,583	4.2%
Liaoning	37%	38%	39%	19,624	5.1%
Ningxia AR	28%	41%	34%	21,239	4.3%
Qinghai	28%	43%	36%	22,679	3.9%
Shaanxi	24%	39%	34%	16,918	4.0%
Shandong	32%	38%	32%	19,228	3.3%
Shanghai	45%	38%	36%	41,188	4.4%
Shanxi	30%	39%	31%	18,300	3.2%
Sichuan	31%	51%	38%	17,852	4.5%
Tianjin	38%	36%	35%	28,682	3.6%
Tibet AR	25%	48%	50%	31,518	–
Xinjiang AR	25%	40%	35%	17,819	3.9%
Yunnan	25%	49%	42%	18,711	4.3%
Zhejiang	40%	37%	33%	27,820	3.5%
Hong Kong SAR	–	–	–	–	–
Macao SAR	–	–	–	–	–

 Sources: **Cols 16, 17** *China Statistical Yearbook 2007*; **Col 18** *China Statistical Yearbook 2008*

19 Health		20 Number of private passenger cars per 10,000 people *2007*	21 Cell phone subscribers as % of population *2006*	22 Internet subscribers as % of population *2006*	Provinces
Doctors per 10,000 people *2006*	**Expenditure on tobacco** as % of household spending *2006*				
35	**1.9%**	**175**	**35%**	**10%**	**China**
28	3.2%	68	20%	6%	Anhui
80	1.1%	1,221	99%	30%	Beijing
28	2.4%	87	38%	8%	Chongqing
30	1.5%	158	43%	15%	Fuijan
33	2.0%	45	21%	6%	Gansu
36	0.9%	316	76%	20%	Guangdong
28	0.9%	76	26%	8%	Guangxi AR
22	3.4%	72	17%	4%	Guizhou
37	1.5%	104	29%	14%	Hainan
34	1.9%	225	33%	9%	Hebei
40	1.2%	129	33%	10%	Heilongjiang
32	1.6%	111	25%	6%	Henan
38	2.6%	87	30%	9%	Hubei
32	2.2%	80	24%	6%	Hunan
43	1.7%	203	36%	7%	Inner Mongolia AR
36	2.2%	230	38%	14%	Jiangsu
28	1.9%	52	22%	7%	Jiangxi
47	1.5%	165	42%	10%	Jilin
51	2.5%	183	38%	11%	Liaoning
39	2.0%	123	36%	7%	Ningxia AR
37	1.9%	95	31%	7%	Qinghai
37	1.8%	128	32%	11%	Shaanxi
36	1.0%	219	31%	12%	Shandong
60	2.0%	330	89%	28%	Shanghai
44	2.1%	218	29%	11%	Shanxi
29	2.3%	124	24%	8%	Sichuan
58	1.5%	525	56%		Tianjin
32	5.8%	125	22%	24%	Tibet AR
49	1.3%	134	33%	6%	Xinjiang AR
27	4.6%	138	24%	8%	Yunnan
43	2.8%	352	60%	6%	Zhejiang
17	–	–	–	61%	Hong Kong SAR
–	–	–	–	–	Macao SAR

Sources: **Cols 19, 21, 22** *China Statistical Yearbook 2007*; **Col 20** *China Statistical Yearbook 2008*

Provinces	23 Water resources		24 Water use		25 Industrial waste water discharged meeting standard
	Total available *2006* km³	**Per person** *2006* m³	**Total** *2006* km³	**Per capita** *2006* m³	*2006* 10,000 tons
China	**2,533.0**	**1,932**	**579.5**	**442**	**2,178,461**
Anhui	58.1	949	24.2	396	68,097
Beijing	2.2	142	3.4	220	10,098
Chongqing	38.0	1,357	7.3	261	81,146
Fuijan	162.3	4,578	18.7	528	124,960
Ganxu	18.5	710	12.2	471	13,103
Guangdong	221.6	2,396	45.9	497	199,215
Guangxi AR	188.1	4,011	31.4	670	119,795
Guizhou	81.5	2,176	10.0	267	10,006
Hainan	22.8	2,735	4.6	558	6,956
Hebei	10.7	156	20.4	297	121,750
Heilongjiang	72.8	1,905	28.6	749	39,344
Henan	32.2	343	22.7	242	121,024
Hubei	64.0	1,122	25.9	454	82,930
Hunan	177.0	2,795	32.8	517	91,618
Inner Mongolia AR	41.1	1,720	17.9	747	21,416
Jiangsu	40.4	538	54.6	727	280,457
Jiangxi	163.0	3,769	20.6	476	59,739
Jilin	35.4	1,300	10.3	378	32,010
Liaoning	26.1	616	14.1	333	88,007
Ningxia AR	1.1	177	7.8	1,294	11,980
Qinghai	56.9	10,431	3.2	590	3,487
Shaanxi	27.6	739	8.4	226	36,118
Shandong	19.9	215	22.6	243	141,540
Shanghai	2.8	154	11.9	660	47,146
Shanxi	8.9	263	5.9	176	30,377
Sichuan	186.6	2,278	21.5	263	97,456
Tianjin	1.0	95	2.3	217	22,925
Tibet AR	415.7	149,001	3.5	1,255	223
Xinjiang AR	95.3	4,695	51.3	2,529	12,556
Yunnan	171.2	3,832	14.5	324	30,568
Zhejiang	90.4	1,829	20.8	422	172,414
Hong Kong SAR	–	–	–	–	–
Macao SAR	–	–	–	–	–

 Sources: **Cols 23, 24, 25** *China Statistical Yearbook 2007*

CHINA: NATURAL RESOURCES 中

26 Energy production			27 Electricity consumption 2006 billion kwh	28 Industrial waste gas emission 2006 1,000 tons	Provinces
Coal production 2006 million tons	Crude oil production 2006 million tons	Natural gas production 2006 million m³			
2,373	**185**	**58,553**	**-**	**44,856**	**China**
83	0	0	66	1,303	Anhui
7	0	0	61	256	Beijing
40	0	647	41	1,274	Chongqing
19	0	0	87	790	Fuijan
40	1	153	54	866	Ganxu
0	13	4,895	300	1,827	Guangdong
7	0	0	58	1,913	Guangxi AR
118	0	0	58	1,882	Guizhou
0	0	205	10	45	Hainan
84	6	655	173	2,915	Hebei
103	43	2,465	60	1,182	Heilongjiang
195	5	1,868	152	2,984	Henan
11	1	116	88	1,394	Hubei
59	0	0	77	2,159	Hunan
298	0	0	88	2,487	Inner Mongolia AR
30	2	61	257	2,035	Jiangsu
28	0	0	45	1,212	Jiangxi
30	7	241	41	953	Jilin
74	12	1,194	123	2,393	Liaoning
33	0	0	38	584	Ningxia AR
7	2	2,503	24	291	Qinghai
183	20	8,047	58	1,633	Shaanxi
141	28	855	227	2,869	Shandong
0	0	564	99	631	Shanghai
581	0	602	110	3,182	Shanxi
86	0	15,995	106	2,222	Sichuan
0	19	1,050	43	344	Tianjin
0	0	0	–	4	Tibet AR
43	25	16,420	36	1,005	Xinjiang AR
73	0	17	65	917	Yunnan
0	0	0	191	1,285	Zhejiang
–	–	–	–	–	Hong Kong SAR
–	–	–	–	–	Macao SAR

Sources: **Cols 26, 27, 28** *China Statistical Yearbook 2007*

西部大开发
中国电信

COMMENTARIES ON THE MAPS 

1 TRADE

It may be many years before China challenges the USA's global military dominance. Trade, however, is another matter, as China has become a trading superpower. Taking Hong Kong into account, China ranks second in the world, and is predicted to overtake the USA by 2035. China is also expected to dislodge Japan as the world's second largest economy by 2020. As a major trading nation, China exerts considerable power and influence in the global arena, but even more so in the region where the USA has reigned supreme since World War Two.

There has been a remarkable growth in trade since China's open policy was declared in 1978 – a far cry from the previous Maoist program of striving for self-sufficiency. That policy was not without its success, for economic growth grew at an annual rate of 6 percent from 1952 to 1978. But foreign trade grew at a slower rate, with China's leaders fearful of losing control of the nation's industry to foreign powers. Another dramatic shift has been towards the export of manufactured goods. This cannot be credited solely to the availability of cheap labor, but also to sound economic management, which has led to striking improvements in the quality of Chinese goods and reliability of delivery, thus making them competitive on the world market. China's trade, however, is more than seeking markets for its products or investment in the economy; crucially, it also needs to import oil, and to procure military and civilian high technology.

It is still too soon to measure with any degree of certainty the impact of China's accession to the World Trade Organization. Membership opens foreign markets to goods from China, and in return opens China's domestic market to foreign competition. This has proved problematic as China has accumulated a huge trade surplus, straining relations with its foreign competitors.

There is a further issue, however, as to what extent China can rely on trade as its springboard to world power. There are challenges and even threats to its domestic tranquility and stability, as related elsewhere in the atlas. These include a shaky financial and banking system, corruption, growing inequalities, inflation, unemployment and severe environmental problems. A growing pernicious nationalism may strain relations with its partners, fuel hot-headed adventurism overseas. Relationships with the USA may erupt over Taiwan or, more likely, when America's status as the world's only superpower is challenged. Nor is China secure from vicissitudes of the global market.

Sources: Kynge J. *China Shakes the World*. London: Weidenfeld and Nicolson, 2006 • Buzan B and Foot R, *Does China matter? A reassessment*. London & New York: RoutledgeCurzon, 2004.

2 INVESTMENT

Foreign direct investment (FDI) has been the motor of China's economic growth. China offers plentiful cheap labor, available land, minimal restrictions on working conditions, an authoritarian government committed to market-led reforms, and a huge domestic market. In turn, foreign firms create employment, supply expertise and training, and bring technology. China gains economic growth and makes new friends. Business executives can be counted as among China's best allies. They persuade governments to restrain their criticisms of China's human rights abuses and lobby for export liberalization.

The impact of this partnership on China's economic development is staggering. According to the report World Investment Prospects to 2011, China is a leading destination for FDI, and outstrips other developing countries.

China's response has been the overdue investment in infrastructure. Foreign investors have responded by seeking control of their joint enterprises or by setting up wholly controlled subsidiaries. Foreign companies manufacture two-thirds of China's exports and, even worse, these are mainly assembly plants, reprocessing imports for the market abroad. In the pre-Reform period this was what the Maoists feared and what stimulated their drive towards self-sufficiency. It is not difficult to imagine what they would think of the WTO. China's membership of the WTO allows foreign investors to enter new fields, including telecommunications, banking, insurance and the securities market. While attention focuses on China's ability to attract investment for both the domestic and export markets, it should be remembered that China is also a direct investor abroad.

The concentration of FDI in the eastern coastal provinces, has at least two important consequences. First, there has been an increase in the power of these provinces relative to the center, or, stated differently, there has been a *de facto* devolution of political power from central government to the provinces. Second, the disparities between the coastal and inland provinces are substantial. The Party-State has addressed the first through the reform of taxation policy and the second through its "Go West" program which initiates its own significant investment as well as attracting foreign investment for the central and western provinces. Development has been uneven, with government investment devoted to infrastructure in the hope that Chinese and foreign investors would follow.

Perhaps, more important, is China's ability to maintain political and social stability in order to attract and keep foreign investors. Reports of unrest and local conflicts, however disparate, are not reassuring for the investor. Unfettered growth is not an iron-clad guarantee of political stability in an authoritarian Party-State.

Sources: *World investment prospects to 2011*. London, New York, Hong Kong: Economist Intelligence Unit,

2007 www.eiu.com • Kynge J. *China shakes the world.* London: Weidenfeld and Nicolson, 2006 • Harris S. China in the global economy. In: Buzan B and Foot R, *Does China matter?* op. cit. • Nolan P. *China at the crossroads.* Cambridge: Polity Press, 2004.

3 MILITARY POWER

China is unlikely to come anywhere near challenging the USA's military dominance in the near future. The USA's military expenditure is in a league of its own. Yet China is determined to increase and prove its military capacity, and believes it has good reasons for doing so in regard to both its external and internal security needs. The military is also an important and powerful political actor, however wedded and beholden to the Party. The PLA priorities are clear. The increase in military spending is not only to meet the demands of the global war on terrorism, but for more advanced equipment and high technology weaponry, and for salary increases and better conditions for PLA personnel as well. The testing of a direct-asset anti-satellite weapon and the launching of its first lunar orbiter suggest more ambitious plans.

The 11 major industrial groups that make up China's defense industry have been instructed to make a great technological leap forward to catch up with the USA. Inadvertently, the EU arms' sale embargo has forced China to develop its own systems and to do so quickly. Russia is the main alternative source of equipment and technology, and China is its largest customer. China, however, strives to achieve self-sufficiency in arms' production, so for every Russian system purchased there is a Chinese system being developed.

In increasing its contribution to the UN Peacekeeping Operations, China is facing up to the international war on terrorism. China further makes it clear that its support for international cooperation is conditional on acting through the UN, based on "conclusive evidence", "clear targets" and the "norms of international law". Attention may focus on China's peaceful rise and economic development but this should not detract from its growing military power.

Sources: Smith D. *The state of the world atlas,* London: Earthscan and Berkeley: University of California Press, 2008 • IISS. *The military balance 2008* Oxford: Oxford University Press, 2008 • Smith D. *The state of war and peace atlas.* London: Earthscan 2003 • Shambaugh D. *Modernizing China's military.* Berkeley and London: University of California Press, 2004.

4 INTERNATIONAL RELATIONS

China may not be a world power yet, but it is beginning to behave like one. As one of the world's largest economies, it has reason to do so. The USA views China as a potential rival, and exerts pressure through both competition and control. And as China has sought acceptance as a member of the global community, the USA has withdrawn towards unilateralism.

China is engaged in a series of international political, economic and strategic relations, both bilateral and multilateral, with its close neighbors and with those further afield, including many countries in Africa. Since 1997, Taiwan's influence in the continent has declined, and the mainland has made its presence felt, powering ahead with infrastructure programs of significant benefit to developing communities, but often relying on Chinese labor, and allowing cheap textile imports from China, to the detriment of African workers and local industries. The 2006 China–Africa Summit marked an important stage in this process, and the Forum on China-Africa Co-operation (FOCAC) is the multi-national policy face of a complex set of bilateral relationships between African nations, Chinese provinces and the national government in Beijing. The West is nervous of this burgeoning transcontinental friendship and, as with all neo-colonial narratives, this is one to watch.

China has been integrated into the global community through its membership of international institutions, as signatory to multilateral treaties and as a contributor to UN peacekeeping operations. Perhaps most important is its membership of the WTO. After 15 years of negotiations, and acceptance of many of the demands for reducing tarriff barriers, China gained membership in December 2001.

China's development strategy of a market-led economy and political stability is driven by international trade. Acceptance into the WTO legitimates this. It also helps legitimate the WTO, and for that matter, other international institutions that cannot maintain their status and power without the membership of the world's largest country. The significance of China's membership depends on whether it meets it human rights obligations, and whether its decision makers can balance domestic modernization with international ambitions.

Sources: Alden C. *China in Africa.* London: Zed Books, 2008 • *The new great walls. A guide to China's overseas dam industry.* International Rivers Network. 2008 July. www.internationalrivers.org • Chinese in the Pacific: where to now? The Australian National University. CSCSD Occasional Papers (1), 2007 May • Foot R. *Rights beyond borders,* Oxford: OUP, 2000 • Nolan P. *China at the crossroads,* Cambridge: Polity Press, 2004 • Saich T. *Governance and politics of China,* Basingstoke and New York: Palgrave Macmillan, 2004.

5 CHINESE DIASPORA

When the Dutch/Indonesian/Australian/Chinese media scholar Ien Ang first wrote her now famous thoughts on "On Not Speaking Chinese", she was referring to the anomalies of being, or being perceived to be, of a particular ethnicity and culture, whatever one's linguistic profile and whatever the depths of one's roots beyond a particular nation-state, in this case, China. Her work still serves to remind us that the Chinese diaspora involves multiple migrations, and many quite different stories, outcomes and identities.

The Chinese diaspora is calculated at more than 30 million people worldwide. Some of those included in this figure are long-standing citizens and residents in every corner of the globe, whilst others are just starting on the journey of discovery. Recent migrants from the Chinese mainland and Taiwan to the Pacific island nations are an indicator of those two entities' attempts to create zones of influence in the area. Older migrant communities on the islands trace their roots back to ships crews who were plying their trade independently of indirect governmental directives.

The reasons for migration are as diverse as the motivations of other migrations the world has seen: trading opportunities overseas, poverty, lack of opportunity at home, educational opportunities overseas, political agreement and disagreement, and repression. The Singapore-based scholar Wang Gungwu gives four main formations of migratory pattern in and from the Chinese world. The worker (*huagong*) or "coolie" – a pejorative word usually applied to indentured labor – has contemporary resonances with unskilled migration and people smuggling. The trader (*huashang*) is a common figure of both historical and modern migration, and a role that enables skilled and semi-skilled small business people to extend their interests or shift their opportunities to new markets. The category of sojourner or settler (*huaqiao*) is a catch-all term that really applies now to all Chinese overseas, and suggests an ongoing connection between the Chinese diaspora and the Chinese nation through a kind of overseas patriotism. Finally, there is the multiple migrant (*huayi*), often a cosmopolitan professional, highly skilled, and successful and who may not consider themselves related in any significant way to "China". But of course, levels of education are not the only indicator of long-term cultural separation from the Motherland. Nor is it true to say that educated migrants have left China behind them. In recent years, there has been substantial investment from the diaspora, both in terms of finance and in terms of personal and strategic involvement in China's reform policies and the rebuilding of the national economy.

Allen PM. Contemporary literature from the Chinese "diaspora" in Indonesia. *Asian Ethnicity*, 2003. 4(3). pp. 383-99 • Ang I. To be or not to be Chinese: diaspora, culture and postmodern ethnicity. *Asian Journal of Social Sciences*, 1993. 21(1). pp. 1-17 • Ang I. *On not speaking Chinese: living between Asia and the West*. London: Routledge, 2001 • Wang G. *China and the Chinese overseas*. Times Academic Press, 1991 • Hearn A. *China and Latin America: the alliance of the 21st century*. Durham NC: Duke University Press (forthcoming 2009) • Research School of Asia and the Pacific, Australian National University. Chinese in the Pacific: Where to now? CSCSP Working Papers 1. May 2007. Canberra: ANU.

6 POPULATION

In the 1990s, when the first edition of the atlas was conceived, the population problem was still a metonymic shorthand for thinking about China. By 2007, the rate of population growth had slowed, although the actual population was still large, and is still growing.

From the sheer weight of numbers, policy implications flow. In 2007, 8.1 percent of the population of China was aged over 65 and this proportion is expected to rise to 22 percent, or 320 million people by 2040. Although this is a measure of the improvements in lifestyle, medicine and nutrition, it presents the problem of how China is going to deal with the pensions, housing and health care needed on such a massive scale. And, to complete the picture, the declining birth rate means there will be a smaller workforce to support a larger elderly population.

Uneven distribution of the population is a reflection of economic growth, and developmental contradictions emerge when the labor force is both a power and a problem of the reform era. China's leaders have pursued "a highly interventionist and aggressive policy", whereby "population planning has shifted from being pro-natalist in the 1950s to being anti-natalist". Population control is seen as an imperative to modernization, even though it conflicted with the belief in the 1980s and 1990s that it is China's cheap and plentiful labor force that makes it competitive in a global market. However, there is already significant unemployment and, following its entry into the World Trade Organization, and more recently the downturn in the global economy, China has faced shrinkage in both light and heavy industries, and more lay-offs to manage among the workforce. The modernizing population has middle-class, aspirational consumption patterns which support capital growth. This increases as urbanization progresses, but many people will be left out of the lifestyle explosion.

Furthermore, China's capacity to operate as the production house for many international concerns is likely over time to become less crucial than its ability to offer outsourcing for highly trained knowledge workers. Again, the increased population in the western regions may not be able to find work, whilst only those with access to good education and training in the east will do well – but they will then bear a burden of welfare for the aged and the unemployed.

Sources: *Women and men in China. Facts and figures 2007.* Department of Social, Science and Technology. National Bureau of Statistics, 2008 • Conway G. *The double green revolution.* London: Penguin Books, 1997 • UN Population Division statistics http://esa.un.org/unpp/

7 THE GENDER GAP

Gender relations in China have been characterized by an historical contempt for women. Even in recent times, when the educational levels and employment achievements of the intellectual elites at least equals those of sisters in developed nations, women continue to be hampered by the priority given to men in social and cultural organizations. In rural areas, the situation is significantly worse, and recent analyses of the last census show a correlation between gendered work on the land and poor levels of education. This is happening particularly in those agricultural areas otherwise described as developed, "because in the agricultural sector of these zones essentially women and older people are working" (National Bureau of Statistics, 2004).

Work is also a relevant way of plotting new class relationships in a modern society. Female migrant workers are ubiquitous in cities, and they are usually confined to low-paid domestic labor or factory work. At the other end of the spectrum, however, a new entrepreneurial class of women – already seen in educated elites in Shanghai and Beijing, is also emerging in central provinces. Despite the status of so-called "non-working wives" (*meigongzuo de laopo*), women are intrinsically involved with business, and their work is of central importance to the development of the family's entrepreneurial interests. In the public sector workforce, women still tend to occupy informal, part-time, and low-wage positions. Women who appear in services statistics as "professional and technical" are generally nurses or primary teachers. These roles are respected, but carry less status and lower pay than many of the male-dominated professions. In periods of unemployment, women endure longer gaps between jobs due to social prejudice against married workers, and a general preference to see men in work. Figures show that there has been little change over the past five years in the service sector in regard to the types of roles undertaken by women, a disappointing statistic.

Although the importance of family life to a healthy nation is a common thread running through the 20th century – in the social policy of the Imperial, Republican and Communist regimes – there have been exceptional periods. In the late 1960s, girls were encouraged by political idealism to put off marriage and sex, and in the 1970s, the birth-control program placed an emphasis on the way people behaved in relation to the rest of the population, rather than in relation to their own sexuality or desire for a family. This policy was called the later-longer-fewer (*wan-xi-shao*) plan. Women would marry later, have longer spaces between births, and fewer children overall. In 1979, the one-child policy was announced: a system of rewards and fines, designed to significantly decrease the proportion of children to adults. In 1981, the policy was tightened to demand that all families should restrict themselves to a single child until the year 2000. The plan was that the size of the population would stabilize in the interval, without having a long-term effect on the composition of society. Unfortunately, the fundamental problem of gender inequality has led to abusive practices against unwanted girl children, and latterly to abduction of prospective wives. The policy was reviewed in 2006–07, but it will take years for the gender imbalance to work its way through the system.

A paucity of women of marriageable age has not curbed the tendency towards domestic violence in the home, and it has been estimated that 30 percent of Chinese families, a figure involving 80 million domestic units, are scenes of violent domestic crime of this sort. The All Women's Federation highlights these cases in media reports, and academic women's studies groups and NGOs are also active in working to change policing and public attitudes in this domain. There are also attempts to publicize the violence meted out to women trafficked for sex, often without their knowledge or consent, and to canvas the ways in which they should be protected from abuse.

Neither the Cultural Revolution of the 1960s, nor the delay of marriage in the 1970s had overturned the basic social assumption of male superiority, nor of women's imperative to be mothers and wives. There has been progress since Liberation (*jiefang*) in 1949, however. Women are now involved in public life, especially at local levels. The number of female deputies in the National People's Congress increased from 147 in 1954 to 626 in 1993, and has remained steady at approximately 21 percent of delegates in subsequent meetings. In 2008, the Vice President of the All Women's Federation pointed out that this figure was unacceptably low, especially given higher percentages of involvement at lower levels – 48 percent on urban residents' committees for example. Many political women are furthermore engaged in, or assigned to "women's issues", a situation which exacerbates gender divisions in the sphere of government. They do not receive the backing they need from male mentors, and the glass ceiling is still hard to break through. The highest female achiever to date (with the exception of Mao's wife, Jiang Qing) at least in terms of government position has been the redoubtable Wu Yi. She retired in 2008.

The future of Chinese women is mixed. If the gender gap continues to widen there will be more violence against women as they become a scarce, but undervalued, "human commodity" in an increasingly market-oriented society. Some women will benefit from their single-child status, but this is not a significant factor, as intellectual and professional families have long taken the education of their daughters very seriously. Women will suffer if the employment rate drops, and men demand access to jobs in a shrinking field of opportunity. The gender gap is a symptom rather than a cause of women's struggles for equal status with men in public, economic and domestic life in China today. Women will need to increase their role in the building of governmental systems, in accessing economic power in the era of reform, and in avoiding the worst degradations of capitalism on their bodies and self-respect.

Sources: Du F, Yang J-C, Dong X-Y. Why do women have longer unemployment durations than men in post-restructuring urban China? July 2007. PMMA Working Paper No.23. Available at SSRN: http://ssrn.com • *South China Morning Post* at www.scmp.com, various issues • Jeffreys E. Over my dead body! Media constructions of forced prostitution in the People's Republic of China. *Portal: Journal of Multidisciplinary International Studies*, 2006. 3(2) (online) • Pun N. *Made in China: women factory workers in a global workplace.* Duke University Press, 2005 • *Far Eastern Economic Review*, various issues • McLaren AE, editor. *Chinese women – living and working.* London: RoutledgeCurzon, 2004 • National Bureau of Statistics www.stats.gov.cn/english

8 MINORITY NATIONALITIES

Although minority nationalities in China make up a small proportion of the population, because of the size of the overall population they comprise a very large number of people. Just one minority group, the Zhuang, are as numerous as the current population of Chile, and the minority nationalities as a whole outnumber the population of Australia by a factor of six. The term, "minority nationality" (*zhongguo shaoshu minzu*), is based on a Stalinist definition of nationality: "an historically constituted community of people having a common territory, a common language, a common economic life, and a common psychological make-up which expresses itself in a common culture." (Wang, 1998) In theory, this definition works in favor of political citizenship, against a pan-Chinese ethnic identity. In practice, there are documented economic and social disadvantages attached to minority status for many groups.

Recognition of minority status currently applies to 55 groups, with the dominant Han majority making a total of 56 officially recognized groups. Despite the insistence on discrete national identities, there are many similar elements within the experience of many minority groups, whatever their "nationality". These can be summed up as: economic disadvantage, poor representation at national levels, religious repression and educational disadvantage. Direct repression is only encountered when the State perceives religion operating as a galvanizing force in demands for independence (as opposed to autonomy).

The most highly publicized case is that of Tibetan Buddhism, and there were serious riots and demonstrations by young Tibetans, including monks, in March 2008. However, the situation in Xinjiang is, if anything, more troubling to central government. Xinjiang (which translates literally and tellingly as "New Frontier") is known to Muslims in Central Asia as East Turkestan, a name that relates the predominantly Muslim Uygur population to their neighbors in the newly independent states of Kyrgyzstan, Kazakhstan, Uzbekistan, Tajikistan, and Turkmenistan. These were formerly part of the Soviet Union, but are recalled in the memory of their indigenous peoples as parts of a greater Turkestan, which converted to Islam 1,200 years ago. The separatists cite religious and ethnic differences for their cause. Xinjiang separatists were blamed for bomb threats made before and during the Beijing Olympics in 2008. It is arguable that the problems may be as much to do with central government containment strategies as with long-standing religious allegiance.

Large-scale migration of Han and other ethnic groups into Tibet and Xinjiang has produced a two-tier economic system, with the Tibetans and the Uygurs sitting at the bottom of the economic ladder in their own areas. Many of the incomers have been troops, stationed semi-permanently to reclaim wasteland and establish new oases. This policy, which echoes age-old centralist ideas of "strengthening frontiers through people" (*yimin shibian*), is dangerous to fragile environments but is also perilous for the people who find themselves washed osmotically out of the structure of development by better-educated and more highly skilled migrants. By 2000, 41 percent of inhabitants of Xinjiang were Han. That figure has remained stable until 2008. The tensions that have arisen globally since September 11, 2001 and the War on Terror, are played out in China's relationship to the East Turkestan Independence movement and other organizations committed to Uighur autonomy. The level and type of their actual threat is difficult to gauge, as support for Islamic movements is muted in western governments and media, and only reported in China when deemed politic to do so.

The dual-structure of the economy of Xinjiang, developed through Han and other ethnic immigration, is structured around modern industry in the north, and an agricultural base in the south. The Uygurs tend to reside in the poorer south, whereas recent migrant populations operate and expand in the north. There have been similar experiences in Tibet and Inner Mongolia, although not as obviously as in Xinjiang. Between 1964 and 1994, 70 percent of migrants into Tibet were themselves Tibetans from nearby provinces, although that situation has changed recently, and the new train line into the area will increase Han access even further. Migrants into Inner Mongolia have been predominantly Han.

Sources: press sources 2000–08 • Mackerras C. *The new Cambridge handbook of contemporary China*, Chapter 9. Cambridge: Cambridge University Press, 2001 • Wang D. Han Migration and Social Changes in Xinjiang. *Issues and Studies*, 1998 July. 34(7) pp. 33–61 • Zhang T, Huang R, *Zhongguo shaoshu minzu renkou diaocha yanjiu* (*Surveys and research into China's minority populations*). 1996 Mar. Gaodeng Jiaoyu Chubanshe • Teheri A. The Chinese Muslims of Xinjiang. *Arab View* www.arab.net • Promoting three basic freedoms: freedom of association assembly and expression, 1997 Sept. www.igc.apc.org/hric • Barnette R, editor. *Resistance and reform in Tibet.* London: Hurst, 1994.

9 RURAL–URBAN INEQUALITY

Inequality in China is spatial, and refers mainly to the origin and location of people and groups. It is also historical and political. People experience very different lifestyles, according to the region in which they live and whether they are urban or rural. The western provinces are rich in mineral resources and space, but historically poor in political influence and basic infrastructure. The coastal provinces have easy access to trade by sea, a strongly developed urban network, and close connections with the seat of power in Beijing.

There are pockets of severe inequality in the east and well as in the less-developed western regions. Anhui is an example. Small and mainly rural, it loses population to the cities on the coast, and lacks investment as there are preferred opportunities for investors in Guangdong and Shanghai. Guangxi, in the southwest, is also isolated, and has suffered badly from floods, high inflation and a decline in rural income levels. Generally, though, it is large, predominantly rural, provinces, such as Qinghai, Gansu, Yunnan, and Guizhou that are most disadvantaged by their distance from the center, by their low proportion of successful commercial activities, and by the slower rate of modernization they experience.

The government is working hard to develop these regions, and the attention to rural education costs and infrastructure is an indicator of their resolve. However the campaign to "open up the west", which was again emphasized in Wen Jiabao's 2008 policy speech to the National Assembly, creates problems as well as benefits. Classic tactics for economic development include: inward migration of skilled migrants, education for all local children and young adults, investment in telecommunications, electricity, and roads, and fiscal decentralisazation.

Many of the poorer rural regions in China are populated by minority nationalities (pp. 30–31). Unrest in these areas is usually blamed on terrorism, religion and separatists movements. However, it is just as arguable that internal disaffection in rural areas is caused by economic disadvantage, which minorities blame on Han inwards migration. Thus, one strategy undercuts the others, as newly educated children cannot develop their skills in the workplace and return wealth to the household, and a spiral of poverty persists amongst indigenous peoples in the area. The decision in 2008 to allow land reform and land leasing by peasants may go some way to alleviate at least one cause of rural disaffection. However, given the power of developers and large businesses to accumulate property, it will be necessary for peasant-farmers to work together to insist on the long-term value of their land before they agree a price.

A major policy response is to devolve financial and fiscal powers to the provinces (*fangquan rangli*) in the hope of controlling resources, containing unrest, and developing economic links across China that are perceived to be equal and mutually profitable. The distribution of wealth across the provinces is complicated by the large income differentials within provinces. Rural counties, even those adjacent to one another, are experiencing varying levels of economic advantage. This may be another incentive to increasing provincial autonomy, which goes hand in hand with existing attempts to "match up" rich and poor provinces in an effort at poverty alleviation. This is known as "horizontal regional co-operation" (*hengxiang jingji lianxi*). All such plans, as well as fundamental attempts to reduce illiteracy and safeguard welfare provision are reliant on the continuing solvency of the nation's economy.

Sources: Lin T, Zhuang J, Yarcia D, Lin F, Income inequality in the People's Republic of China and its decomposition 2000-2004. *Asian Development Review*, 2008. 25(1) and (2). pp. 119-36 • Storrocks A and Wang G. Spatial decomposition of inequality. *Journal of Economic Geography*, 2005. 5(1). pp. 59-81 • Fan S, Kanbur R, Zhang X, Regional inequality in China: an overview. 2008 Aug. www.cornell.edu • Tsai DH-A. Regional inequality and financial decentralization in mainland China. *Issues and Studies*, 1996 May. pp. 40–71 • Andreosso-O'Callaghan B and Qian W. The PRC's economy – From fragmentation to harmonization? 1997. unpublished MS • Chang Y-C, The financial autonomy of provincial governments in mainland China and its effects. *Issues and Studies*, 1996 Mar. 32(3). pp. 78–95 • Fu F-C and Li C-K. Disparities in mainland China's regional economic development and their implications for central–local economic relations. *Issues and Studies*, 1996 Nov. 32(11). pp. 1–30 • Liu AP. Beijing and the provinces: different constructions of national development. *Issues and Studies*, 1996 Aug. 32(8). pp. 28–53.

10 ECONOMIC DEVELOPMENT

According to the IMF, in 2007 China overtook the USA as the world's most powerful engine of growth. Its Gross Domestic Product, which admittedly started from a low base, averaged 10 percent annual growth from 1990 to mid-2008. And even following the downturn of the world economy, it is likely that China's economy will retain its global importance.

China's growth has been largely based on high investment (both foreign and domestic) and on its strong export markets. Initially, the impact of the country's economic development was apparent on the eastern seaboard, but since 2002 the central provinces have seen the fastest rate of change. In the worsening global economic climate of 2008, Wen Jiabao urged the expansion of domestic consumer demand, in order to make the country less reliant on overseas markets.

Commentators have pointed to various factors behind China' economic growth, including ambitious government targets and the ideology of endeavor that drives the Chinese people to meet them. But the over-riding factor is that the Party-State's ruling legitimacy rests largely on its ability to deliver a healthy economy, and improve the general conditions of the Chinese people. Only in this way is it likely to maintain social stability.

Source: Wen JB. Government work report to NPC 2008. http://npc.people.com.cn • Huang Y. *Capitalism with Chinese characteristics: entrepreneurship and the state.* Cambridge: Cambridge University Press, 2008 • Kynge J. *China shakes the world.* London: Weidenfeld and Nicolson, 2006 • Harris S. China in the global economy. In: Buzan B and Foot R. *Does China matter?* op. cit. • Nolan P. *China at the crossroads.* Cambridge: Polity Press, 2004.

11 ENTREPRENEURS

The new face of state capitalism in China, and the emphasis on creating a market economy, has seen a rapid increase in private enterprise, and in the number of entrepreneurs. The growing entrepreneurial culture is also a symptom of improving ties between mainland China and "Greater China" – the term used for Chinese populations in Hong Kong and Southeast Asia that have a developed commercial connection with the People's Republic. Whatever political differences are at play, Chinese populations continue to consider one another as natural and viable partners in business and trade. These relationships have matured over the past ten years, and have been encouraged by China's accession to the WTO, and by the more local CEPA agreement between Hong Kong and the mainland – generally understood as a local Pearl River Delta initiative with Guangdong province. The idea of "Greater China" is regional and strategic, frequently drawing together a triangle of southern China, Taiwan and Hong Kong–Macau. This informal grouping is developing in parallel with international organizations, such as ASEAN (which excludes Taiwan but is traditionally anti-Communist), and APEC (Asia-Pacific Economic Cooperation), which incorporates non-Asian countries into a forum for inter-regional co-operation.

Entrepreneurialism at home responds to these large opportunities from overseas and to domestic conditions. State entrepreneurialism is evident at city and provincial levels, and also across certain sectors, such as education, advertising, real-estate development, and tourism. The Shanghai Television University, for example, has developed distance and online learning in order to capture the vocational market from its rivals in the tertiary sector, and other television channels. As Jane Duckett pointed out, state entrepreneurialism is likely to grow with the availability of markets, and the improvement in bankruptcy laws and loans systems. A political will to devolve economic micro-management allows state actors to take more responsibilities and more risks in their sector or region.

Individual entrepreneurs are operating within a similar set of factors. The de-collectivization of state assets, the reform of land use and tenure, migration and urbanization, increasing unemployment, and the need to look after one's own and one's family's welfare in old age, is pushing people into ambitious ventures. Again, quoting Duckett: "Entrepreneurialism may have spread because of perceptions that everyone else is doing business, and those who do not feel that they are missing opportunities."

Sources Goodman DSG. *The new rich in China.* London: Routledge, 2008 • Jaffrelot C and van de Veer P. *Patterns of middle class consumption in India and China.* London: Sage, 2008 • Krug B, editor. *China's rational entrepreneurs: the development of the new private business sector,* London: RoutledgeCurzon, 2004 • Donald SH, Flew T, and Wang X. "The Administration of creativity: China. Leadership and the new MBA" (unpublished) • Duckett J. *The entrepreneurial state in China,* London: Routledge, 1998 • Goodman DSG, Why women count: Chinese women and the leadership of reform. In: Anne E. Mclaren, editor. *Chinese Women: living and working.* RoutledgeCurzon: London, 2004 • Blecher M, Benewick B, Cook S. editors. *Asian politics in development: Essays in honour of Gordon White.* London: Frank Cass, 2003.

12 EMPLOYMENT

The nature and requirements of work in China are very different from the years of peasant agriculture and heavy industry, supplemented by political and intellectual work. People have re-invented themselves as migrant laborers, service workers, small-scale entrepreneurs, and business managers. These re-inventions are in many cases linked to higher incomes, and more ambitious career options, and an increase in professional management posts. Although many university graduates are still assigned to public posts in the state sector, employment now is seen as a personal responsibility. Even public servants need to think about their entrepreneurial skills in order to keep their jobs.

There are several factors contributing to new employment profiles in China: reform and modernization, recent history, and regional imbalance in capital distribution and labor costs. Unemployment surfaced in the late 1970s when tens of thousands of people in their twenties and early thirties returned to the cities from the countryside. These were "urblings", city youth who had spent much of their life since the age of 17 working in remote rural areas as part of the management of the Cultural Revolution. Their impact on employment figures were, however, minimal in comparison with the influx of migrant peasants which they foreshadowed. And the effects of surplus labor continue. According to official statistics, 26 million people were laid off from the state-owned enterprises between 1998 and 2002 and more followed as China needed to meet commitments under the WTO agreement. Some of the short-term pain has been directed to the private sector. The policy of "holding onto the big ones but letting the little ones go" (*zhuada, fangxiao*) ensures that only strategic, larger enterprises are kept under the wing of total state ownership.

There remain deep regional divisions. Employment patterns in the coastal regions are capital intensive and highly mechanized; laborers supplement their incomes with alternative and seasonal employment. Meanwhile, farm practices in remote regions are labor intensive but generate very low wages. Family members, often young women, need to move into towns and cities as guest workers to supplement the household income. They achieve a certain status through their shift into economic power, but their situation in the urban centers is often precarious. Development in industrial practices often, at least in the short term, leads to downscaling, re-organization, and the subsequent displacement and reconfiguration of the workforce. Add to this that many employment opportunities are off-set by the mass population growth.

The downturn in the global economy has resulted in factory closures and in increasing unemployment in the Pearl River Delta and the eastern coastal provinces. This impacts on the inland labor-exporting regions, such as Chongqing, which sustains a population of over 30 million. Its municipal government predicts that 180,000 migrants could return over the next four years and is helping them find work and set up new enterprises locally.

Women workers are particularly disadvantaged by the likelihood of unemployment. Over half of newly laid-off workers are women, although they make up only 45 percent of the workforce. This adds to an already weak position for women. They are further challenged by the illegal, cheap labor on hand from poorer regions. The choice is being made between social problems and a strong national economy. It is still uncertain which way the state will move in the long term.

Sources: Migrants start to move back inland. *China Labour Bulletin.* 2008 Oct 28 • Migrant remittances in China: the distribution of economic benefits and social costs. In: Murphy R, editor. *Labour migration and social development in contemporary China.* London: Routledge,

2008 • *China labour statistical yearbook 2007*, China Statistical Publishing House 2007 • Jacka T. *Rural women in urban China: gender, migration, and social change.* London and New York: ME Sharpe, 2006 • Chow PCY. China's sustainable development in global perspective. *Journal of Asian and African Studies.* 38(4-5), 2003.

13 AGRICULTURE

Every change, however small, in feeding one-third of the world's population impacts on the rest of the world. The great success of China's reforms has been to lift 200 million people out of poverty and to introduce a degree of farmers' participation in village decision making.

The Party-State, however, has still to resolve crucial problems of Chinese agricultural and world development. The impressive gains in production as a result of the re-introduction of household farming in the early 1980s reached a plateau and has since stagnated. Although the conditions of farmers' lives have improved the income gap between the cities and countryside has widened. Environmental abuses, corruption, and the misappropriation of land, have ignited considerable unrest.

The Party-State has set out the ideological framework, social harmony and scientific development, for tackling these problems. For example, agricultural taxes have been abolished, and new measures are being introduced which will allow farmers to directly transfer their land. If successful, farmers will receive compensation for the land lost, while farming will be introduced on a larger and theoretically more efficient scale. This will encourage more factory farming, already estimated at 64,000 farms, compared with 18,800 in the USA, and perhaps more adulterated products.

Sources: MacDonald M. Opinion: Chinese farms a growing challenge. Worldwatch Institute. 2008 Oct 20 • China's government helps boost grain output, 2004 March 4. http://english.people.com.cn• East Asia Analytical Unit, DFAT, Australia.

14 INDUSTRY

Industry in China is in a state of flux. In the 2003 edition of this atlas we showed that industrial output was booming. The state sector, however, was in trouble and the momentum for growth was shifting to the private sector. Since then, state-owned industries have experienced something of a renaissance. The combined profit of the 150 companies controlled by central government rose by 223 percent between 2003 and 2008. More dramatically, four of the 10 most valuable companies in the world at the end of 2007 were Chinese state-controlled, although the valuation reflects the Shanghai stock market. There is a growing mix between state and private companies, and even a willingness of state-owned enterprises to compete in highly competitive industries.

China's manufacturing, however, is not immune from the global economy. Exporters have been hit by rising costs, stronger currency, product counterfeiting and contamination, and the global crises. It has been reported that 67,000 small firms in the first six months of 2008 in the Pearl River Delta in south China have gone bust. This is the region of China well-known internationally for the export of toys, shoes and electronic products, produced mainly by migrant labor. Dongguan City's 14 million inhabitants are almost migrant workers from the countryside.

Sources: Dyer G and McGregor R. China's champions. Why state ownership is no longer proving a dead hand. Financial Times online. 2008 Mar 16 • Branigan T. Reverberations of world recession rock a city built on exports. *The Guardian.* 2008 Nov 1 • Kynge J. *China shakes the world.* London: Weidenfeld and Nicolson, 2006 • Jefferson G et al. Ownership, performance and innovation in China's large and medium-size industrial enterprise sector. *China Economic Review*, 2003. 14(1). pp. 89-114.

15 SERVICES

The service sector was a political non-starter in the years before the reform era of the 1980s and 1990s. Maoist perspectives emphasized production, and the only valid services were those provided by government agencies. The re-growth of service industries, particularly in southern China, has moved in fits and starts since the 1980s, and with varying degrees of transparency. Department stores, luxury goods, up-market restaurants, and late-night clubs are features of most cities. By 2006, 11,828 hotels and 1,332,083 rooms had been registered by the China National Tourism Administration, a tenfold increase in 12 years. Real-estate, information and finance markets, retail and personal services have all grown fast – although the downturn in the international financial sector will mean some tightening of belts as investors re-evaluate their priorities. The service sector as a share of GDP in 2006 was 70 percent in Beijing, 50 percent in Shanghai and 40 percent across the whole of China, and there is little household debt, so the sector still has room for growth.

There is also some overlap with what is known as the "quaternary sector" – digital and information-based services across the economy. ICTs are important to all parts of the service economy, and well-designed and maintained ICT networks are crucial to business and public infrastructure. This includes tourism and e-commerce, health and education, finance and governance. The growth of the ICT professional sector and the research and development needed to create home-grown companies and locally relevant models is hampered by piracy, however. In 2008, Microsoft tried to re-instate its IP on pirated software across China, and was met with a barrage of criticism by users, who did not appreciate black screens. Nonetheless, the fact remains that they were using stolen IP and that the impact of piracy in China not only affects costs for users elsewhere, it also prevents the local ICT industry from developing niche markets.

Unplanned and statistically invisible services are also on the rise. The domestic sector is booming, with most professional women with children employing a nanny, maid or *baonu* to do the domestic chores. Hawkers are also common on city streets, but they need to be licensed. Those who come in from the countryside to sell off surplus produce have operated under "specialized agricultural households" licenses. Those selling food or manufactured produce on city streets are also required to obtain "temporary business licenses". The penalty for non-compliance can be fairly straightforward. Police are quite likely to throw large piles of unlicensed produce into the canal.

Sources: *China Development Brief*, various issues • Law P-L and Chu W-CR. ICTS and China: an introduction. *Knowledge, Technology and Policy*, 2008 Mar. 21(1). pp. 3-7 • Ma JX, Buhalis D, Song H. ICTs and internet adoption in China's tourism industry. University of Surrey e-press, 2003 • Donald SH, personal interviews • White LT III. *Unstately power: local causes of China's economic reforms*. New York: M E Sharpe, 1998.

16 TOURISM

The tourist industry is a major growth area in the Chinese service sector. Although pilgrimages and festival days have promoted internal travel across most periods of Chinese history, modern tourism is a recent development. Thee most up-market expansion is still in the major cities. In 2007, there were 131 million tourist arrivals, 26 percent of whom were foreigners. The majority of these were near neighbors: Japanese, South Koreans and Russians. Nonetheless, 28 percent of foreign receipts come from long-haul air travel.

Some companies are centered on villages lucky enough to be situated near a genuine "site". Villages in rural Beijing at the foot of the Great Wall moved fast in the 1980s to develop service facilities (and hawking) to maximize the benefit of their proximity. Between 1949 and 1972 the wall was only accessible to "special friends of the Revolution" and only 250,000 visas were issued in those years. The wall is now a necessary part of any Beijing visitor's itinerary.

Tourism benefits are measured by receipts and by the impact on development. The phenomenon of local site development is now happening all over China. Guizhou, just one of many possible examples, has invented traditions to bring together its real and imagined minority cultures into a package for tourists. Unfortunately, this sometimes leads to local inhabitants moving out of the area as entrepreneurs move in and change the economic balance of real-estate and labor needs. Even eco-tourism can have this unfortunate side-effect and thereby worsen conditions for the rural poor. If well managed, however, the rural citizen can find a new role in managing the local political and social environment through collaborative tourism activity.

Despite the long-running international discomfort with China's presence in Tibet, there is a strong focus on the region to attract domestic (Han Chinese), and some international, tourists. Lhasa is becoming the fulcrum of tourist routes across the plateau, and the region is investing heavily in rail, roads and facilities to continue the trend. Chinese tourists are also venturing further afield, accounting for 34 percent of all arrivals across Asia.

Hong Kong has long enjoyed the status of prime destination in the region, but in recent years the city has been struggling to retain that pre-eminence. Strong branding campaigns to promote Hong Kong as "Asia's World City" have been directed both at international and mainland markets since 2000. In October 2004, 164,000 mainland visitors travelled to Hong Kong in the first three days of the Golden Week holiday. Golden Week is a vacation break invented to coincide with the lunar festival, and to encourage domestic tourism around China. By 2007, 55 percent of all visitors to Hong Kong came from the People's Republic – an increase of 12 percent from 2006.

Sources: Ryan C and Huimin G, Perceptions of Chinese hotels. *Cornell Hotel and Restaurant Administration Quarterly*, 2007. 48. pp. 380-91 • CNTA, various bulletins • Oakes T. Cultural strategies of development: implications for village governance in China. *The Pacific Review*, 2006 19(1). pp. 13-37 • World Expo and the new development in Shanghai: 2010 Expo and Shanghai tourism. In: Shanghai: Shanghai Tourism and Enterprise Committee, Policy and Regulation Section, 2003. p.12.

17 ENERGY

The 11th National Plan includes two key quantitative targets that are somewhat contradictory. Firstly, China aims to achieve an annual gross domestic product (GDP) growth rate of 7.5 percent, with the goal of doubling 2000 GDP per capita by 2010; but secondly, China also wants to reduce energy consumption per unit of GDP by 20 percent, and the total discharge of major pollutants by 10 percent, by 2010. The only way to achieve both these objectives is to invest in innovative green technology, to implement anti-polluting measures across industry and service sectors, and to find ways to contract over-consumption without retreating on growth altogether.

China has had such success in the recent past. Between 1978 and 2000 there was an improvement of two-thirds in the ratio of energy used against units of production. This was possible because of the very poor levels of energy management in the pre-Reform era. From 2001 to 2006 energy demand grew four times faster than predicted on the 2000 figures, and China was creating 15 percent of global demand. The problems are twofold. Demand is centered on heavy industry. In 2005, 70 percent of energy was used by industry, 10 percent was residential usage, 7 percent was transportation, and 13 percent other sectoral uses. However, the much-needed greening of industry is slowed by intra-provincial competition for quick growth. Whilst consumer usage is not nearly as high per capita as it is in the USA, nonetheless China's ecological footprint is creeping higher against its overall bio-capacity (2.1:0.9), when compared to its developing neighbor India (0.9:0.4). Consumer usage is therefore a major challenge over coming years. China's ongoing subsidies for electricity and petrol are unhelpful in training consumers to prioritize environmental design choices in housing and transport. It is predicted that China will account for 20 percent of world energy demand by 2030.

Supply is also a problem. The infrastructure for delivery is neither secure nor efficient. This is a hangover from old public systems, but also falls short in the realm of ICTs and integrated governance. The crucial question for China is whether its energy resources are sufficient to meet its growing needs. Natural gas, which makes up 3 percent of China's energy production, is located in the remoter parts of China. Australian LNG is piped in to make up the shortfall. Oil makes a major contribution to the country's energy needs, but despite being the sixth-largest oil producer in the world, China is now a net importer. Both off-shore production and oil-field explorations in Xinjiang province have proved disappointing, and China has resorted to buying oil fields in countries such as Azerbaijan and Indonesia. Explorations in Africa are also linked to the need for diverse energy supplies in a period of international instability. Hydroelectric power is also significant because it can be created locally – another factor in the relentless dam construction in China.

Sources: Wen JB. Government work report to the National People's Congress, Beijing 2008. http://npc.people.com.cn • WHO Country Profile, 2008 • Rosen DH and Houser T. *China energy: a guide for the perplexed.* Center for Strategic and International Studies and Peterson Institute for International Economics, 2007 • *China Daily*, various issues www.chinadaily.com.cn • *Nuclear power in China.* World Nuclear Association, 2004 Sept. www.world-nuclear.org

18 URBANIZATION

At the beginning of the reform period in 1978, 81 percent of China's population was classified as rural. Today, the proportion of the population that is rural is down to 55 percent, following the global trend towards urbanization.

Even where the population has grown in rural areas, this picture is complicated by the number of towns and small cities that have sprung up in those regions, and by the vast extent of domestic migration both within and across provincial boundaries. With the re-designation of towns as cities, and the economic status of cities on the rise in every province, the rural resident who does not have dealings in the urban environment is increasingly unusual. The flow of domestic migration is very great; it is estimated that several million (perhaps as many as 200 million) people are moving constantly between places of work and rural places of origin. In 2006, $270 billion, or 23 percent of China's fixed investment spending was used to build the necessary real-estate and infrastructures to make space for in-bound migrants. Many, of course, are not adequately housed and still live in urban "villages", which are broken up whenever the city government so chooses. By June 2005, 86.7 million migrants were registered as temporary city dwellers, and millions more were unregistered. A registered migrant can purchase a land lease or home, and thereby get an urban passport (blue stamp *hukou*), if they have sufficient cash. This is unusual, however.

Rural job seekers are replacing urban workers by undercutting the market, and are causing long-term employment problems. Since the migrants are non-permanent residents, still in many cases without access to infrastructure and basic amenities, city populations are often underestimated – and under managed. Although migration does not necessarily entail movement to the biggest metropolitan districts, but often a lower scale of intra-provincial movement, the continuing intensity of population density on the eastern seaboard means that China's population remains unevenly spread across its territory. Companies also complain that their workers leave without warning to return home, so that urban development is seen as uneven and problematically dependent on outsiders (*waidiren*). By 2005, 11 provinces (although not the big receivers such as Guangdong and Beijing) has relaxed their *hukou* policy to encourage continuity.

China's cities are most usefully understood as administrative centers with a very great concentration of people. City administrative districts can be visualized as in concentric circles: the old city/inner city; suburban districts which contain district-administered towns; commune/market towns. Then, outside the urban area, but within the municipality, there are rural counties that contain county-administered towns and more market towns and communes. Cities include the province-level cities of Beijing, Shanghai, Tianjin and, most recently, Chongqing, which is sometimes described as the largest city in the world; capitals of provinces or autonomous regions, or state-planning cities such as Dalian (Liaoning) and Qingdao (Shandong Province), which are responsible to the central government. For the purposes of investment and development some, such as Dalian and Qingdao, are also designated as Open Cities, and a smaller number, such as Shenzhen, are Special Economic Zones. The 2003 trade agreement (CEPA) between Hong Kong and mainland China encourages cross-investment and production. It is likely that such links will foster major urbanization, as the conurbation of Hong Kong's new territories leaps the border to stretch right across the Pearl Delta and Guangzhou.

Sources: Ma LJC. Urban administrative restructuring, changing scale relations and local economic development in China. *Political Geography*, 2005. 24(4). pp.477-97 • Cartier C. City-space: scale relations and China's spatial administrative hierarchy. In: Ma LJC and Wu F editors. *Restructuring the Chinese city: changing society, economy and space.* London: Routledge, 2005. pp. 21-38 • Danwei.org (various issues) • Benewick R. Towards a developmental theory of constitutionalism: the Chinese care. *Government and Opposition*, Autumn 1998 • Ma Xiaohe, quoted from: Tianze Institute report. China reports 8 percent GDP growth in 2003. *People's Daily Online* http://english.people.com.cn accessed 2002 Nov 5 • China migration country study. Ping H. Institute of Sociology, Chinese Academy of Social Sciences, Beijing and Pieke FN. Institute for Chinese Studies, University of Oxford, UK. Presented at the Conference on Migration, Development and Pro-Poor Policy Choices in Asia. Dhaka, 2003 June 22-24. Published on internet • China's urban population to reach 800 to 900 million by 2020. *People's Daily*, 2004 Sept 17. cited on China Economic Net.

19 TRANSPORT

In 1952, 67 percent of travelers in China journeyed long distances by rail. In 2007, only 6 percent of all journeys were made by rail (although with only a fractional decrease in actual numbers travelling, owing to the massive increase in total population). Over the same period, highway traffic took up the dominant position, moving from 19 percent to 92 percent of domestic volume.

Motorized transport in China is still predominantly communal – buses and taxis are much more common than private cars – but that situation is changing rapidly. There is also a vast regional variation in the provision of public transport. In Beijing in 2002, there were 823,800 buses, cars and trucks on the roads. In Henan, a province with approximately seven times the population of Beijing, there were only 691,500. Overall, the situation is one of rapid increase. In 2007, there were 37 million vehicles in China, with a predicted number of 370 million by 2030. 5 million new cars were sold in 2006 alone.

Road traffic is still diverse in China, with trucks, pedal-cabs, taxis and sleek personal cars vying with the many bicyclists, almost none of whom wear helmets. It is arguably this diversity that causes many of the road traffic accidents in major cities, and makes travelling by road in the countryside and the city outskirts so hazardous. There were 104,372 deaths attributed to traffic accidents in 2003, and 81,649 in 2006, 7,806 fewer than in 2005. The downwards trend is encouraging, although China still has the highest death-toll per accident, in the world. Car use is being monitored in highly polluted cities, but petrol is still very cheap. On the bright side, however, Shanghai is also building a public-transport hub to support the 2010 World Expo, and in preparation for the 2008 Olympics Beijing completed four new subway lines, and has seven more in planning and construction.

Long-distance travel is crucial to the opening up of the Western Provinces and to economic restructuring in general. The rail link to Tibet is a case in point, although a sore point for proponents of less Han influence and more independence. Air travel is increasing sharply, in response to intra-provincial business opportunities, domestic tourism and international travel. The lower-paid workers, described as domestic migrants, will, however, still be more likely to travel by train.

Sources: China Safety Forum, 2008 • World Health Organization press release announcing publication of *World report for road traffic injury prevention*, 2004 Oct 8 • BBC News online, 2004 Sept 3 • Harvard China Project. Urban transport, land use, air quality and health in Chengdu • The road to progress. *Asiaweek,* 1997 May 9. p.10 • *China Statistical Yearbook*, various years • Thanks to Guo Yang (Shanghai Tourism Commission); ISTP, Murdoch University.

20 AIR POLLUTION

In February 2007, Linfen in Shanxi was dubbed the world's most polluted city. Host and hostage to coalmines, cooling furnaces and iron foundries, Linfen is one of the powerhouses of a province which is doing relatively well under industrial reform. But the price is too high. Across China as a whole there is at least one city in every province (except Tibet and Guangxi Autonomous Regions), which exceeds the WHO interim target for polluted developing nations. Air pollution on this scale is a major threat to health in China. Children are described as living in an atmosphere that produces effects equivalent to smoking two packs of cigarettes a day. Related health damage as a proportion of GDP was 1.7 percent in 2000, and is predicted to reach 2.5 percent in 2010, and 3.3 percent in 2020. The effects include premature mortality, chronic bronchitis, and fatal asthma attacks.

Pollution is increased by an over-capacity of fossil-fuel power plants, even in areas expecting to use hydroelectric power in the near future. One explanation for this over capacity is localized short-term productivity gains. The power plants look good, on paper at least, in comparison to other smaller industrial enterprises. They thereby enhance the political reputation and financial standing of local officials.

A casual regard for air quality results in part from the anarchy in nationwide electricity controls. Coal-smoke is the major pollutant in Chinese cities, causing smog in certain weather conditions. Outside the cities, the effects of fossil-fuel use are exacerbated by a loss of forest cover. Forests in Sichuan and Jiangsu have been decimated since the 1950s, and forest cover along the Yangtze dropped by over 50 percent in the last 30 years of the 20th century.

Rural and urban indoor pollution is also noted by Chinese scientists. The main causes are fossil-fuel stoves (some without flues), and inadequate ventilation, especially in winter. These dangers acutely affect women working at home during the day. One can compare the situation

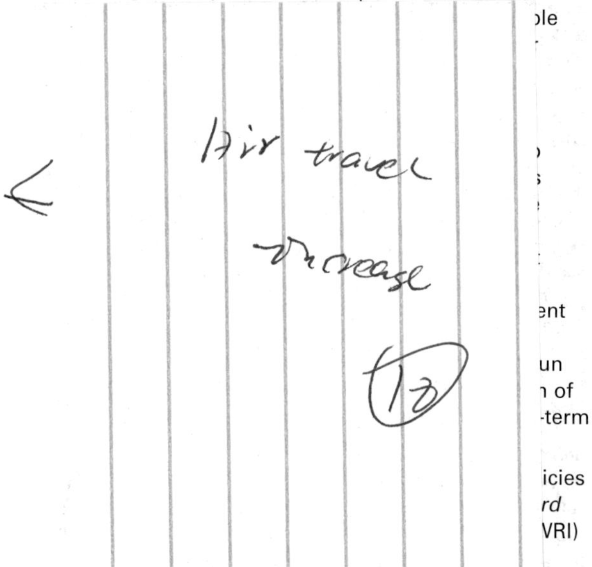

benefits of Greenhouse Gas Mitigation Policies in China. Environment for Development discussion paper, 2008 Apr • Ho MS and Neilsen CP editors. *Clearing the air: the health and economic damages of air pollution in China*, MIT Press, 2007 • Weller RP. *Discovering nature: globalization and environmental culture in China and Taiwan.* Cambridge: Cambridge University Press, 2006 • Fenby J. *Observer*, 2004 Aug 18 • China Environment Forum at the Woodrow Wilson International Center for Scholars www.wri.org/wr-98-99/prc-ntro.htm • Smith Kirk R, Gu S, Kun H, Qiu D, One hundred million improved cookstoves in China: How was it done?. *World Development*, 1993. 21(6). pp. 941–61 • *Shanghai Newsletter*, 1997 Aug 9 http://www.shanghai-ed.com

21 WATER RESOURCES

China's water is unevenly distributed, and some would argue that climate change is making weather patterns less predictable. Seasonal flooding appears to be getting more extreme, and droughts more frequent and more severe. In 2007, 177 million people were affected by flooding, 1,230 people died, and over 1 million houses collapsed. Flood-control measures cost China 34 percent of its annual investment in water resource and management. Some areas were first affected by drought then devastated by floodwaters and mudslides. The common aspect is a flooded south and a dry north. Desertification caused by bad land management is a particular problem, contributing to air pollution as well as water seepage and waste.

The monumental Three Gorges dam was developed to deal with flooding problems, facilitate water-borne transport systems, and provide hydroelectric power, but has been hugely controversial. Questions have been raised about the migration and resettlement of local populations, the quality of the project design, the eventual cost of power, but above all experts have questioned whether it will be effective in controlling the vast power of the Yangtze. The dam is a byword for the water conflicts that are reported (and rumored) to take place in China, and which also simmer at its borders. Dam construction dislocates vast communities and local activists contest new dams and new disruption; upstream polluters cause tension for downstream users; and – on the international scale – China can access, and pollute, the origins of its neighbors' water sources (such as the Mekong River). All of these problems need to be addressed by inclusive strategies for management, grievance-resolution systems, and relevant technology.

In the plains of the northeast, agriculture and industry make heavy demands on available surface water, to the extent that the Yellow River regularly runs dry before reaching the sea. Irrigation is using increasing amounts of water (an 18 percent increase from 1990 to 2002), and critics point to the excessive wastage involved. An increasing reliance is being placed on underground water, which is being used at an unsustainable rate. In some places the water table is dropping by 3 meters a year, and in 2007 the total water resources fell by 2.5 percent on 2006.

An ambitious "solution" has been devised: the South–North Diversion Project. Although this major engineering feat will, when completed in 2050, result in the diversion of a flow of water similar to that of the Rhine, in Germany, it will still only meet a small proportion of the north's water deficit. In addition, the water will be used, legitimately and illegally, along the way; there is expected to be a high rate of evaporation, and levels of pollutants will increase during the transportation. In 2007, alone the project cost 8.7 billion yuan,

China's water use is in line with the way water is used in the rest of the world with a slightly higher emphasis on industry: 81 percent is industrial use, 8 percent is domestic and 1.8 percent is environmental flow. Domestic use does, however, vary widely between provinces, and between rural and urban areas. In most provinces urban residents use significantly more water in their home than do their rural neighbors. This is probably because access to tap water is more common in urban areas than rural, but lack of information on access to tap water among rural households makes this difficult to verify. Government statistics claim an access rate of 40 percent of the rural population, leaving 60 percent of rural residents without running tap water. In urban areas, claims are complicated by the reality of shared households, in old blocks, where people living on higher storeys do not have taps on their floors, in which case "access" is actually severely curtailed in practice.

Sources: Ministry of Water Resources, *Statistical bulletin on China water activities*, 2007 www.mwr.gov.cn • Sampat P. Groundwater mining. WorldWatch Institute, 2000 • South-to-North water diversion project, China www.water-technology.net • Wilson W. China Environment Series, 2005-08.

22 WHO RULES CHINA

In Part 5 we have abandoned the conventional approach to the Party-State in order to show more clearly who exercises supreme power. In doing so, we offer a different and awesome reality, that the destiny of one-fifth of the world's population is ultimately in the hands of nine men. Not only do they occupy the commanding positions of the largest political party in the world, but the commanding positions in the State Council, National People's Council and the Military. This approach is also intended to give a clearer picture of just how pervasive and dominant the Communist Party is.

Power structures are complex, and that of China's Party-State is no exception. A closer examination reveals, firstly, that China's government can be seen to possess those functions commonly associated with a political system: executive, legislative and judicial. Secondly, it is possible to identify those characteristics that are special to China, in particular, the Standing Committee of the National People's Congress (NPC) and the State and Party's Central Military Commissions. Thirdly, while the state constitution specifies the National People's Congress as the highest organ of state power it is replaced here by the State Council in respect to the exercise of power. Fourthly, rather than a separation of powers or functions, the branches of government are shown to be linked and in an order according to the power they exercise. Fifthly, China is a unitary state in that there is no formal division of powers between the center and provinces, and that the structure of the Communist Party parallels that of the state, extending from the center of power to the grassroots, where Village Committees and Community Residents Committees are elected throughout China.

This does not mean that other institutions are unimportant. The centre and the provinces, for example, compete for resources, as do the provinces with each other. The constitution designates the National People's Congress as the highest organ of state power. In respect to decision-making, our approach suggests otherwise, and most commentators focus on what it does not do, rather than what it does do. They tend to dismiss the NPC as a rubber-stamp legislature that meets once a year to hear the reports of the state (and party) leaders. Certainly, the indirect election of its 3,000 members undermines its authority and independence.

Since the mid-1980s, however, the NPC has been attempting to assert its authority. First, its cause has been championed by successive chairmen of the Standing Committee, appointed from the upper echelons of the Party-State power structure during this period. The present chairman, Wu Bangguo, is in a position to further or to curtail this process. Secondly, the economic reforms have been generating a growing volume of legislation, which is being met by a rise in professionalism and institutional resources. Thirdly, there have been changes in legislative behavior in so far as a sizeable number of delegates are beginning to propose amendments to legislation, delay legislation, withhold unanimous approval of appointments and work reports, and introduce their own items for consideration. Fourthly, there is evidence of lobbying mainly by the delegates on behalf of the local interest. Fifthly, the Standing Committee, which meets on a regular basis between the annual sessions of the NPC, is a working body. When the performance of China's incipient legislature is compared with that of more mature legislatures, the development of the NPC stands up reasonably well. When compared with the State Council it is clearly subordinate, but this is not out of line with executive–legislative relations elsewhere.

Sources: Fewsmith J. *China since Tiananmen. From Deng Xiaoping to Hu Jintao.* Cambridge: Cambridge University Press, 2nd ed. 2008 • Pan PP. *Out of Mao's shadow: the struggle for the soul of new China.* London and New York: Simon & Schuster, 2008 • Blecher M. *China against the tides*, London and New York: Continuum, second edition, 2003 • Saich T. Reform and the role of the state in China. In: Benewick R et al. editors. *Asian Politics in Development*, London and Portland, Oregon: Frank Cass, 2003 • Burns JP. Governance and civil service reform. In: Howell J, editor. *Governance in China*, Lanham, Maryland and Oxford: Rowman and Littlefield, 2004 • Shambaugh C, editor. *The modern Chinese state*, New York and Cambridge: Cambridge University Press, 2000.

23 CHINESE COMMUNIST PARTY

China is governed by a relatively small number of leaders, who exercise power, both formally through a multiplicity of power structures, and informally through a network of contacts. Viewed from this perspective, China does not differ from most nation-states. Yet it is radically different. As the most powerful of the few surviving communist states, the Party leadership refuses to entertain the possibility of a legitimated and institutionalized opposition. Even the Democratic Parties remain under the leadership of the Communist Party.

China's formal power structures include, first and foremost, the Communist Party. Other power structures include the government, the bureaucracy, the PLA, the judicial system including the police, and the provinces individually and collectively, but all of these power structures, although separate, are pre-eminently related to the Party. The structures of the Party parallel and penetrate those of the government, with Party Leading Groups functioning at all levels of the state, and there is an overlap of personnel. The Party contends and proposes, the state amends and disposes.

The formal power structure, how power is exercised informally, and who exercises it can all be described in terms of a pyramid structure. This pyramid structure is also evident in the organization of the Party. There is a hierarchical relationship, concentrating power at the apex of the pyramid and exercising control down to the base. Democratic centralism may be the form providing the opportunities and avenues for debate and discussion at all levels of the Party to be communicated upwards, but decisions are transmitted downwards.

Decision-making is concentrated in the Standing Committee of the Politburo, who rule on behalf of the Party. It is informed by a small number of Leading Groups, each responsible for a policy area and headed by a member of the Standing Committee. Power and its exercise, however, is not always visible, and in China much depends on the base of support, personal networks and institutional backing. An analysis of the membership of the Standing Committee demonstrates how power has passed from a generation of revolutionary leaders to a generation of technocratic leaders. These leaders, however, are reaching the end of their tenure in office and will be replaced by those already known as the "fifth generation". Two factions can be identified. The first are those who have advanced through the Communist Youth League, which has connections with the General Secretary of the Party, Hu Jintao. The second faction is known as the "Princelings", who are the children of high officials. The "Princelings" have a strong base in the PLA since their families would be from the revolutionary generation. The two factions are more or less balanced at this stage.

Sources: *Women and men in China. Facts and figures 2007.* Department of Social, Science and Technology. National Bureau of Statistics, 2008 • Lam W. China 2008: changes in the Chinese leadership and Beijing's new policies on reform, Tibet and Taiwan. *China Brief.* Jamestown Foundation. 2008 May 15 • Brodsgaard KE and

Zheng Y, editors. *The Chinese Communist Party in reform.* Routledge, 2006 • Lieberthal K. *Governing China*, New York and London: Palgrave Macmillan, second edition, 2004 • Chu Y-H et al, editors, *The new China.* Norton, 1995, 2004 • Saich T. *Governance and politics in China.* Basingstoke and New York: Palgrave Macmillan, 2004.

24 THE PEOPLE'S LIBERATION ARMY

The state in China is almost always referred to as the Party-State. The structure of the Communist Party parallels that of the government at all levels and crucially dominates the government at each level. An analysis of the Party-State, however, must also include the armed forces, the People's Liberation Army (PLA). This is in part historical, for the PLA was formed in 1927, 22 years before the founding of the People's Republic. It was also formed on the basis of the integration of Party and military leaders. Mao Zedong set out the principle "that the Party commands the gun and the gun must never be allowed to command the Party". The instrument of control is not a state ministry of defense, but the Party's Central Military Commission, which is presided over by the Party General Secretary, Hu Jintao.

The actual relationship between the Party and the PLA is symbiotic and mutually reinforcing. The first two generations of China's leaders derived much of their legitimacy from their participation in the anti-Japanese war and the two civil wars that brought the Communist Party to power. The Party-State relies on the PLA to ensure political stability and to project China as an international power. The PLA has divided loyalties between its historical commitments as a people's army and ambitions to become a modern military force. A modernization program was launched in the mid-1980s to reduce the size of personnel from 4.3 million to 2.25 million.

The Chinese White Paper on National Defense of 1998 stated: "During the new historical period the Chinese Army is working hard to improve its quality and endeavoring to streamline the army the Chinese way, aiming to form a revolutionized, modernized and regularized people's army with Chinese characteristics." This is reflected in the rapid annual growth of defense spending, which showed a whopping 17.8% increase in 2006–07. Subsequent White Papers, confirmed the modernization program, while emphasizing "active defense", referring to the reunification of Taiwan, countering terrorism and stopping separatist movements and safeguarding political and social stability.

Sources: Shambaugh D. China's Communist Party: atrophy and adaptation. Berkeley, California: University of California Press, 2008 • IISS. *The military balance, 2004–05.* Oxford: Oxford University Press, 2008 • Deng Y. *China's struggle for status.* Cambridge, 2008 • White Paper on China's National Defense. Information Office, State Council of the People's Republic of China, 1998 • Fourth White Paper on China's National Defense. Information Office, State Council of the People's Republic of China, 2003 • Shambaugh D. *Modernizing China's military*, Berkeley and London: University of California Press, 2004 • Shambaugh D, editor. *The modern Chinese state*, Cambridge and New York: CUP, 2000.

25 RULE OF LAW

The rule of law was emedded in the constitution in 1999. The first White Paper on the legal system, published in 2008, reveals that the National People's Congress has enacted 229 laws covering the seven branches of legislation, including the constitution and constitution-related laws, civil and commercial laws, administrative laws, economic laws, laws on society, criminal law and litigation and non-litigation procedural laws. In addition, the State Council has enacted nearly 600 administrative regulations, local people's congresses over 7,000 regulations, and autonomous ethnic areas 600 regulations. Rules have also been put into effect at various levels of government. We have detailed this because constitutionalism and law is a progressive and developmental process, suggesting that China has moved from "rule by persons" through "rule by law" on to "rule according to the law". As for the rule of law, the White Paper states it has been enshrined as a fundamental principle and that the country will be governed according to the law. A socialist party will be governed under the rule of law.

Certainly, this body of law is impressive, and important legislation has been enacted, legislation protecting private property rights and those promoting the rights of labor, for example. But the White Paper and that echoed by Prime Minister Wen Jiabao suggests that the rule of law remains aspirational. There is also another body of law for the Communist Party. This is known as the "three top priorities" or the "three supremes", stating that the legal system must give priority to the party's cause, the people's interests and the constitution and the law. This is interpreted as the party's priorities superseding all else. Law in this sense is instrumental and represents a backward step on the path of the rule of law.

This does not bode well for human rights, which have only been guaranteed by the constitution since 2004. Four years later a "National Plan of Action for Human Rights" was announced. While this suggests significant, if uneven, progress, the Party's instrumental view of the legal and court system could well negate this.

Sources: Information Office of the State Council. *China's efforts and achievements in promoting the rule of law.* White Paper. Xinhua 2008 Feb 28 • Benewick R. Towards a developmental theory of constitutionalism: the Chinese case. *Government and Opposition*, Autumn 1998 • Information Office of the State Council and the Ministry of Foreign Affairs. *National plan of action for human rights.* Xinhua. 2008 Nov 4 • Cohen J. Body blow for the judiciary. *South China Morning Post.* 2008 Oct 18 • Ping J. The rule of law progressing in China. 2008 Oct 31. china.org.cn • Lam W. The CCP strengthens control over the judiciary. *China Brief*, Jamestown Foundation, 2008 July 3.

26 STATE VERSUS CITIZENS

The title "State versus Citizens" signifies how difficult it is for China's leadership to translate its own view of human rights as well as its commitment to international standards into legislative provisions and judicial behaviour. Attention is also drawn to the need for the existence of organizations between citizens and the state, but independent of the state.

The explanation lies not only in China's traditional bias in favor of the state, but also in the paradox of the market. The socialist market-led economy is not, as some would believe, a contradiction in terms. Yet the market does shift the emphasis away from collective and towards individual rights and values, and away from self-sufficiency and towards international inter-dependence and co-operation. The market creates interests, domestically and internationally, and these interests demand and, indeed, need to be heard and expect the state to respond. Market behavior extends beyond the economy into the formation of citizens' organizations. In China these may be under the leadership of the Communist Party, state-registered organizations, unofficial bodies, underground organizations or political and social movements.

The market, however, also creates new complexities which may threaten state authority and entrenched interests. This new environment of economic openness can engender political volatility. Rather than risk destabilizing conflicts, the Chinese leadership has chosen to promote political stability and the maintenance of public order. In the absence of institutional channels and recognized procedures, such as an independent judiciary and due process of law, the Party-State has resorted to authoritarian means, undermining both human rights and the market values it seeks. The persecution of the relatively small number of political dissidents and intellectuals is one example. The range of crimes punishable by the death penalty and the number of executions is another.

China's trading partners can overlook or dismiss abuses of international standards of human rights in favor of market access and stability. Yet the paradox of the market holds. If China wants international acceptance it will not be able to continue to defy international standards. Although China by no means stands alone as the only state responsible for human rights abuses, as Amnesty reports make clear, it is cast as a pariah among states, particularly when it suits the interests of nations or interests within those nations. In this context the award by the European Parliament of its prestigious Sakharov Prize for Freedom of Thought to the China human rights activist Hu Jia is well deserved, but not surprising.

Sources: *US Congressional Commission on China, Annual report*, 2007 • *Amnesty International Report 2008. The state of the world's human rights: China.* London: Amnesty International, 2008 • Foot R. *Rights beyond borders.* Oxford University Press, 2000 • Friends of Nature (Beijing). *Green book 2007.* Edited abstract on Chinadialogue, 2008 June 6, online. • Fenby J. China's slow civil awakening. *The Guardian* 2008 Sept 19 • Davis S. Olympic challenges for Chinese grassroots groups. *Anthropology News.* 2007 Dec • Howell J, New directions in civil society. In Howell J, editor. *Governance in China*, Lanham, Maryland: Rowman & Littlefield, 2004 • Du Jie. Gender and governance. In Howell J, editor. *op.cit.* • White G et al, *In search of civil society*, London: Macmillan, 1997.

27 HOUSEHOLDS

"Empty-nesters", people living without any younger relatives, comprised about half of China's 153 million senior citizens in 2008. The key change in household composition in the 21st century is the rise of elderly dependants, and the increase of older people living alone. China's ageing population – 143 million at the national census in 2004, with predictions of 200 million in 2014, 300 million in 2026 and 400 million in 2037 – is the greatest challenge to prosperity and welfare in everyday life. The direction of all Chinese policy is to create a "well off" (*xiaokang*) society by 2020 and to achieve nationwide basic modernization by 2050. An ageing population with inadequate welfare and social security provisions makes this objective very challenging.

Rural areas have a slightly higher percentage of aged people, which is unexpected and problematic given the low incomes of their children. In major cities there is a different concern that wealth in early to middle age might lead to impoverished senior years. Shanghai municipality announced in 2004 that it would take measures to lower the age profile by allowing two-child families for registered residents. The privilege is not extended to rural migrants however, who are perceived as a cause of crime and high population density, without quality (*suzhi*). They will have to return to rural areas to age and die. In 2007, there were 8.6 beds per 1,000 aged persons in care and hospitalization facilities. This will not bridge the shortfall between those with insurance or family support, and those without.

Meanwhile, gender relations in households continue to be difficult for many women. Domestic violence is an acknowledged but unsolved factor in household relations, with an estimated one in five families experiencing violence. In 2002, a survey by the China Law Society revealed that, "12.1 percent say their husbands kick them, 9.7 percent say their husbands throw things, 5.8 percent say they are forced to have sex and 1.7 percent are burnt or scalded with boiling water". Many rural women are thought to "escape" through suicide (proportionally, there are three times as many suicides in the countryside as in cities – arguably due to the vicious cycle of violence and poverty). In urban areas, divorce is on the increase, as is the incidence of single-motherhood in the wake of failed marriages. These social phenomena, perhaps with the exception of high female suicide, are familiar to Westerners, and are accompanied by an equally familiar onus placed on women after the fragmentation of family units.

Sources: China National Committee on Ageing. http://en.cncaprc.gov.cn • Country Statement: China. United Nations Economic and Social Commission for Asia and the Pacific, 2007 • Honig E and Hershatter G. *Personal voices: Chinese women in the 1980s.* Stanford University Press, 1988 • Beaugé F. Women's birth right. *Le Monde Diplomatique*, 1999 Feb. p. 9 • Psychological domestic violence law proposed www.china.org.cn translated by Li Liangdu, 2002 27 Nov • Bezlova A. Population: Shanghai breaks second-child taboo. International Press Service, 2004 Sept 13 • China starts campaigning against domestic violence. ABC online correspondents report. 2004 June www.abc.net.au

28 FOOD

The big questions around food in China are food security and food safety, and the changes in diet – from high fibre to high carbohydrate and high protein, which adversely affect health. Chinese obesity is increasing in step with the changes in diet. Food security refers to the gap between what China produces and what it consumes, and the degree to which it should rely on imports for basic foodstuffs. This matters for poorer rural and working populations whose expenditure on food can get out of control if there is a hike in world prices for grain. The Chinese government claims that it still only relies on imports for 10 percent of its grain supply, and refuses to take any responsibility for food price fluctuations in the world market.

A new range of food production is growing in response to the market economy and its effects on population movement, expectations of lifestyle and income, and the emergence of the entrepreneurial state. The shrinkage of fertile land used for grain is in part due to city-bound migration of peasants, but also because of expanding activities in fruit and vegetable production to support farmers' incomes. These products are immediately profitable to China's agricultural industry, although their growth might threaten food security in the longer term. They also serve international and domestic middle-class dietary requirements, and, post-WTO entry, encourage foreign direct investment in food processing and branding ventures.

The trends in China may be away from a familiar food economy, but the emphasis for such a large, predominantly rural, population must still be the availability and price of grain. An agricultural action plan was announced in March 1999, based on China Agenda 21, which was originally published in 1993, following the 1992 United Nations Conference on Environment and Development. The aim was to achieve sustainable growth in grain production, animal husbandry, fisheries and township enterprises. The first 36 projects included a food-security warning system, water and soil conservation initiatives and animal and plant conservation. The projects calculate that in order to feed a predicted 1.6 billion people in 2030, an annual grain supply of 640 million tons will be needed. The question for China is the degree to which the profits made possible by the WTO should be sought at the expense of grain production and food security on scarce arable land.

Food safety is a serious problem. Supervision and safety mechanisms are in catch-up mode as the population develops new tastes and expectations, and food businesses take up export opportunities and onshore food processing investments. Safety is, for example, related to the swift increase in so-called luxury food production. Fruit and vegetables are cropped several times a year to increase profits, and this is usually achieved through the overuse of chemical pesticides. There is also a danger of pollution in the food chain where low-paid peasant producers deliberately seek to enhance production quickly. The 2008 case of melamine in milk was of this category. The supervision of licenses has shown evidence of incompetence and corruption. In 2007, Zhang Xiaoyu, the former head of the food and drug administration, was executed for corruption. In August 2008, another senior official committed suicide fearing a similar fate, and in September 2008, Li Changjiang stepped down from the same role over the poisoned milk scandal.

Sources: Yue L. Can the US guarantee food safety in China? reprinted danwei.org 2007 May • Walt V. The world's growing food-price crisis. Time online. 2008 Feb 27 • Looking behind the global food crisis. China Economic.net 2008 July 29. http://en.ce.cn • Chern WS. Projecting food demand and agricultural trade in China. *The Asia–Pacific Journal of Economics and Business* 1(1) 1997 • EAC (DFAT, Australia). China embraces the world market, 2003 • China's obesity rate doubles in 10 years to 60 million people health.news.designerz.com • China's changing diet www.iiasa.ac.at

29 HEALTH

The years of famine aside, China under communism has had a good record on health provision. Better results and better services for the majority was the objective of grassroots training and provision. But since the introduction of the market-led economy, the relationship between financial well-being and health has gone into reverse. The symptoms and causes of this decline in provision are manifold. Rural insurance schemes have collapsed due to the demise of collectives and associated co-operative organizations. So-called barefoot doctors no longer receive work points for their service, so now they charge cash. There is a general suspicion of the security of welfare insurance packages, especially in regard to corruption. The State Council's working party on health reported in 2007 that it was shameful that the WHO ranked China fourth from the bottom in a list of 191 countries.

Using the most reliable indicator of general health standards, infant mortality, the World Health Organization finds that there was a vast improvement between 1960 and 1985 (173 deaths per 1,000 live births to 44 per 1,000), which improved further to 2005. although China's own statistics showed a worrying discrepancy between girls babies (27 deaths per 1,000 live births) and boy babies (22 deaths per 1,000 live births). However, the indication is that rural women are still disadvantaged. While the overall percentage of women giving birth in hospital increased from 51 percent in 1990 to 88 percent in 2006, 10 percent fewer rural women had that opportunity, compared with urban women. Homebirths are not necessarily problematic – indeed, they can help women avoid the complications and infections of hospital doctor-led birthing practices – however, proximity to hospital is a usual condition of homebirth in developed countries, especially on first deliveries. Given the two-child rural policy, most of the births recorded in these figures will be first births. However, the experience of SARS in 2003 and an ongoing Hepatitis B and AIDS crisis, has made the government realize, with the prompting of the World Health Organization, that the rural population has dire need of extra resources.

China's health system is threatened most directly by the tobacco industry and by the spread of HIV/AIDS. It is no secret that tobacco companies in developed countries, faced with a significant decline in smoking, anti-tobacco movements, legislation and lawsuits, are targeting the newly industrialized countries. China beckons, with its huge population, market-oriented economy, impressive economic growth rate, rising living standards, difficulties in enforcing legislation and a thriving tobacco culture in which 70 percent of men, and an increasing number of women, smoke.

Health officials are acutely aware of the risks for China's 360 million smokers and the market inroads that foreign brands are making as fashion icons or by being cheap and available as smuggled items. Their position has been strengthened by the publication in 1998 of the biggest study ever undertaken into deaths from tobacco. Scientists from the USA, the UK, and China investigated 1 million deaths and concluded that if the current smoking uptake rates persist in China, tobacco will kill about 100 million males currently under the age of 29. Half of these deaths will occur in middle age and half in old age.

Health officials claim that most of the 1 million mainland Chinese infected by HIV are intravenous drug users. In areas such as Yunnan's border territory with Burma, on the edge of the "Golden Triangle" of Asian drug production, infection rates are estimated at anywhere between 30 and 70 percent. If it is true that drug users are the most vulnerable in China to HIV/AIDS, then the worst-affected area will be in the south. The Health Ministry claims that around 80 percent of users in Guangdong and Guangxi are injecting, compared with 1 percent of users in Shaanxi and Inner Mongolia. However there are also large numbers of infected persons (*ganranzhe*) in other provinces too, notably Henan, the site of the blood donation scandal.

Local government support for Aids orphans and some attention to Aids prevention is improving in areas where local populations, NGOs and government health agencies can work together, In parts of Anhui, there is a prevention and management strategy called "Four (Basic) One's and Three (Life) Lines". The Ones: every village should have one clinic and one tarred road, every family should have a house and a water supply. The Lines: the Civil Affairs bureau should provide monthly subsidies to families who have lost one parent through HIV, more for those who have lost both parents, and free treatment for affected families in local clinics. This pragmatic and localized approach is reminiscent of the barefoot doctor (albeit with expensive drugs and trained personnel) of an earlier era.

An impending threat to health security comes from the ageing population (*see also* 27 Households). These people, and society as a whole, will be severely disadvantaged without the immediate introduction of long-term affordable medical insurance and pension schemes.

Sources: *Women and men in China. Facts and figures 2007.* Department of Social, Science and Technnology, National Bureau of Statistics. 2008. p. 103 • China Development Briefing May 2007 • Avert organization's page on China www.avert.org • Wu Y. Trends and opportunities in China's health care sector. Murdoch University: Asia Research Centre Policy Paper. no 18, 1997 • *Green Book of population and labour.* CASS Beijing • October 1996 Speech by Chen Minzhang (Health Minister). In: Zhongguo xingbing aizhibing fangzhi zazhi. reported on US Embassy Beijing www.redfish.com • Chinese rural dwellers get better medicare service www.chinaview.com 2004 Oct 24 • Chinese Academy of Preventative Medicine et al. Smoking and health in China, 1996 • *National preventative survey of smoking patterns.* Beijing: China Science and Technology Press, 1997 • Zhang J. Cigarette sellers cash in on foreign brands. *China Daily*, 2004 Feb 17 • Global Youth Collaborating Group. Special report: Differences in worldwide tobacco use by gender: findings from the Global Youth Tobacco Survey. *Journal of School Health*, 2003. 73(6): pp. 207-15.

30 EDUCATION

In the first decades of the socialist era education was strongly aligned with ideological training, political development, and the formation of good citizens for the new China. This approach to the upbringing of China's youth was in conformity with general concerns in Chinese society and culture. Education (*jiaoyu*) is an essential component of childhood in Chinese life. Children's films, books, outings, must all have a demonstrable educational and patriotic aspect. But in practical terms, giving a working education to China's children is a difficult task. Private education is one of the solutions chosen by parents with the money to pay for it. The Fourteenth Congress of the Chinese Communist Party (1992), made this easier by its endorsement of "the socialist market economy".

Educational inequalities are deepening in the wake of market reforms. Some parents pay out huge sums for private schools and semi-private institutions (*minban*), and all must contribute to their children's education. In 1956, private schools were abolished as part of the reforms of Liberation. By the end of 1993 there were 125 different kinds of private school in Guangdong. Meanwhile, 80 percent of a total of 1 million primary schools have established enterprises through which to better their financial situation. Other children cannot attend school at all when family farming commitments require their labor, or when finances cannot stretch to the cost of fees and books. A developing trend towards streaming exacerbates the disadvantage of children with interrupted or incomplete schooling. Junior high school students (*chuzhong*) are streamed so that some are moved straight into vocational training whilst others continue into senior high school, and consequently enjoy a more extensive set of options. In order to hit nationwide targets for proportional entry from senior high school to university, however, some education regions have significantly reduced the number of senior high places.

The Education Law requires a standard nine years of compulsory schooling – and policy discussions are moving towards a 12-year target. Hu Angang, a senior researcher at Qinghua University, has argued for the extended schooling range, pointing out that this will develop the nation's long-term human resources and skills base, and will also alleviate immediate pressures on employment: "Estimates have stated that by 2005 over 75 percent of middle-school graduates in urban areas will enter high schools, while the proportion in rural areas will reach 65 percent. By 2010 the percentage in urban areas could increase to almost 100 percent and the figure in rural areas could reach over 75 percent. Such an extension (in schooling) will bring China many benefits, primary amongst which will be its role as an effective measure in diminishing strong employment pressure." Hu's theory will only work if those who deliver education policy at the local level (*fangquan*) are able to fund education. Given the 2008 announcements in March and October that rural education fees would be cut, as far as possible to zero, this might finally come to fruition. In the wake of global economic crisis, the importance of educating China's peasantry (and thus growing their spending power) has become an absolutely crucial aspect of national survival and growth.

The tertiary sector is also growing fast. As an indication of its development, 195,000 overseas students from 188 countries chose to enrol in elite Chinese universities in 2007; this was an increase from 141,000 in 2005. In 2003, there were seven Chinese universities in the top 500 on the Shanghai Jiaotong index, in 2008 there were 18. In contrast to students during the Mao era, contemporary students are looking for vocational, technical and commercial relevance in their education. Sichuan University offers courses in real-estate, marketing, advertising, and interior decoration. Fudan in Shanghai is an elite university which has a growing focus on general business skills: accountancy, enterprise management, municipal planning and economics. These courses are self-financing and very popular. This represents a shift from elite higher education to mass access, and it signals China's knowing embrace of a knowledge economy. However, for the 750,000 Beijing graduates entering the job market in 2007, many of them hoping for white-collar jobs in the city, they may have to opt for rural placements in public-service posts where there is never a queue of returning college graduates to find a first job.

Sources: Wen JB. Government activity report to the National People's Congress, 2008 • Lawrence D, China pledges to eliminate poverty in rural areas. *The Australian*, 2008 Oct 16 • China is popular destination for overseas students. *China Daily*, 2008 Oct 27 • Hu A. 12 years of schooling will benefit entire nation. CERNET, 2001.

31 MEDIA AND TELECOMS

Telecommunications are helping create world communities and intra-nationalal connections at the same scale as cross-border trade, migrations and large-scale interdependencies have done in the past. The change in the nature of globalization lies in the collapse of time and space in the wake of advanced systems of communication.

China has been working in support of a socialist market-led economy to modernize its telecommunications infrastructure systems for over 30 years. Most provincial capitals are now linked by fiber-optic cabling. Mobile telephony is also extremely important, providing a leapfrog technology for rich and poor, rural and urban, and making networking and enterprise start-ups possible for business people in all classes, including the urban and rural working class, and migrants. Indeed, the greatest recent growth in all media enterprises has been in mobile platforms and applications. From 2004 to 2008 these have been SMS, prepaid mobiles, internet cafes and "little smart" (*xiaolingtong*) – an inexpensive mobile phone service that connects via landlines and can only be used within a limited area. This indicates the need for cheap communications for Chinese workers, and the conquest of spatial challenges that such mobility achieves.

The media in China are profoundly important to the operation of business, government and social relations. They are also the bedrock for great creativity, fierce debate and political frustration. As such, they are not unlike the media systems operating across both macro-economies and developing geo-political entities the world over. In China, the contradictions within the media-world have a very special character, however. Media content is tightly controlled through online portal surveillance, censorship and central directives on policy (*tifa*), but media industries are also a powerhouse of profit, talk, debate and emotional expression. A film director may be banned in one context, but be active on television in another. Television serials, game shows and reality formats are very popular, and the more ambitious may tackle serious issues such as divorce, corruption and abduction. In a very few instances, films manage to bridge the gap between entertainment and tough social debate, and still get their films distributed. Jia Zhangke, the director of *Sill Life* and *The World* is a case in point.

Internet access in China is provided through educational and government-sponsored establishments, but internet cafes persist, with strong demand defying the regular clampdowns for security reasons. The internet is monitored for inappropriate and politically sensitive sites and searches. The issue was raised by foreign journalists in 2008 during the run up to the Beijing Olympics, and some restrictions were lifted for certain users in specified sites. Researchers have demonstrated that site blocking can reach 100 percent for certain issues. Domain-name hijacks are sanctioned by government-sponsored servers, usually resulting in unexpected re-routes through cyber space. Human Rights Watch has drawn attention to the imprisonment of those who defy the content restrictions in university chat-rooms. Meanwhile, however, global and domestic IT companies make profits in China, although these are limited by persistent software piracy. IN 2007, the net profit rate for IT companies ranged from 74 percent for Giant, 45 percent for Alibaba (a Chinese company), 36 percent for Baidu (the most popular search engine). And 29 percent for Microsoft. Online users, or "netizens" favor music "mp3" sites, news blogs, and discussion boards. Netizens may offer informed commentary on national affairs, or act as internet vigilantes. The 2008 outings of the addresses and names of so-called Western and Tibetan sympathisers were fuelled by "crowd-based information gathering" (*renruo sousuo*) online.

The internet still poses a challenge to regulation, and the Chinese government has responded with criminal punishments and fines of up to 15,000 yuan (US$1,740) for those who access or provide pornography, anti-government dissent, political propaganda, and amorality. The list is open-ended and gives the authorities plenty of scope for interpretation according to the priorities of the moment. In October 2008, the State Administration for Radio Film and Television shut down 10 sites on the grounds that they carried pornographic (yellow) content. However, one of these sites, Oeeee.com is the community news portal for the bold newspaper *Southern Metropolis Daily*, and is unlikely to have carried anything "yellow". Regulations are also in place for the monitoring and censorship of other media. One cannot publish photographic essays on the lives or work of national figures, alive or dead. Official biographies are authorized, in print and on film, and are tightly monitored for content according to the political line at the time of release. There are also firm controls on the publication and distribution of foreign works, and on the release of foreign movie titles.

Where local content is deemed too boring, especially in the south near Hong Kong and Taiwan, satellites pick up Hong Kong shows. Although in Cantonese, these are more desirable to young putonghua-speaking viewers than Hubei music specials. In 1978, China had 32 television stations and 3 million TV sets. Now, there is reportedly 100 percent penetration and up to 60 free-to-air channels in Chinese households, although that data does not account for villages in remote areas where there is no electricity. Caveats aside, there is significant government commitment to television as a medium, which can be easily monitored – especially through cable delivery. Total digitization of television is planned by 2015. The national broadcaster, CCTV, is now running pay TV channels, and expects to earn more in revenue than it can raise through advertising on its free-to-air services.

Sources: Keane M, *Created in China: the great new leap forward.* London: Routledge, 2007 • Zhu Y. *Television in post-reform China: serial dramas, Confucian leadership and the global television market.* London: Routledge, 2008 • CSM Research Focus 2007 • China Network Information Centre, 2007 • Qiu JL and Cartier C. Networked mobility in urban China: hukou, working class ICTs and the case of Sun Zhigang. Paper presented at the annual meeting of the International Communication Association, TBA, San Francisco, CA, 2008 Oct 23 available from: www.allacademic.com • AsiaMediaNews Update asiamedia@international.ucla.edu • personal interviews with Donald SH, 2006–08 • *Southern Metropolis Daily*, various issues • Media Timeline Danwei.org

32 RELIGION

China is sometimes seen as a secular state with very little patience for superstition and belief. This is both true and profoundly false. Since neologisms to describe the practice of religion, *zongjiao*, and superstition, *minxin*, were introduced in the late 19th and early 20th centuries, there has been a separation between the five "world religions": Catholicism, Protestantism, Islam, Buddhism, and Daoism, and local, grassroots belief systems involving indigenous temples and gods. These latter are "superstitious". Confucianism, and now Neo-Confucianism, has been seen as a case apart, combining as it does the tenets of politics as well as self-cultivation within a social framework.

In mainland China the majority of believers are Daoist (or formally identify as such to avoid the *minxin* label), there are a substantial number of Buddhists, and a growing number of Muslims and Christians.

In all of these *zongjiao* practices, however, it is arguably possible to see the core tenets of Chinese practice being regenerated and pursued: teaching as belief (*jiao*), the worship of Gods and the placation of spirits (both named: *shen*), the recognition of man's place in the cosmos (*tian*). Of course, the differences are also profound. Household gods, protection against ghosts, and ancestor worship (*zu xian*) lie at the core of much religious ritual, although since the founding of the People's Republic in 1949 folk festivals have not been part of the official calendar of national events. Popular Daoism works alongside folk beliefs, local gods and village- or household-centered worship, and is therefore successful and popular. Philosophical Daoism, on the other hand, is concerned with the loss of corporeal signs as motivation in the development of the spirit. Buddhism, and particularly the desire for nirvana, is antithetical to Daoist conceptions of immortality. Christianity and Islam offer a completely different system again, with a focus on a single "true" God, even though each religion disagrees on the nature and commandments of Godliness.

There are other quasi-religious features of Chinese social life, which inflect the basic patterns of belief. Neo-Confucian rationality and quasi-spiritual appeals to harmonization are back in political vogue. The focus on "harmony" (*hexie*) is a continuation of the way in which language was used in the pre-Reform era. Key words of revolutionary discourse, for example, function similarly to the sacred words of religious texts. As Timothy Cheek, a scholar of the revolution, points out "Confucius held that if names were not correct and realities did not conform to correct names, then the moral state would be an impossibility. The Chinese Communist Party exhibits a faith in the power of names similar to that attributed to Confucius". Their use shapes the meaning of everyday life, by giving it an extra dimension, which must be supported by the belief of speaker and, ideally, of listener. It is the communication of belief that occurs in sacred rites and political rallies that brings the concept of religion close to both that of revolution, and to practical political philosophy. However, public religiosity is not allowed to move beyond the limits of religious licenses and patriotism. Hence, local temples sometimes describe their grassroots activities as Daoist to escape censure for superstitious practice, Protestant house churches (arguably the fastest growing network of believers) organize in underground cells rather as the revolutionaries did before them, and Tibetan monks and nuns complain at enforced patriotic education.

The bête noire of Chinese religious politics is indeed the Tibetan question, mainly because it involves not just a challenge to belief in the State, but because it combines religion with threats of separatism. Human rights activists draw on both the right to freedom of religion and the desire for autonomy in their defense of Tibetan independence. The Chinese government argues that religion is not a sufficient reason for dividing national territory. There is also a problem where religion and ethnicity come together in a nexus of traditional racisms. The supporters of Tibet have not been so vocal about similar aspirations amongst Muslim Xinjiang separatists. In that case it seems that Western opinion is more in favor of Chinese national stability.

Interestingly, despite the Chinese government's persistent criticism of the figure of the Dalai Lama as a cipher of separatism, superstition and feudalism, images of Mao Zedong now appear in village temples and as urban fetishes. (Mao's image hangs in taxis for good luck.) Whereas before it was the figure and words of Mao that offered a politico-sacred text to his people, now it is his enshrined memory that encapsulates the ancient and the modern in the Chinese religious consciousness. His nirvana is on the earth that made him.

Sources: Yang MM-H editor. *Chinese religioisities: afflictions of modernity and state formation*. Berkeley: University of California Press, 2008 • Wang Y. University of Sydney/SBS lecture notes • Feuchtwang S. *Popular religion in China: the imperial metaphor*. London: Curzon, 2001 • Cheek T. The names of rectification: notes on the conceptual domains of CCP ideology in the Yan'an rectification movement. In: *Keywords of the Chinese revolution: The language of politics and the politics of language in 20th-century China*. Funded by the National, Endowment for the Humanities and the Pacific Cultural Foundation. www.easc.indiana.edu

SELECT BIBLIOGRAPHY 中

Primary sources

China Daily www.chinadaily.com.cn/english/home/index.html
People's Daily Online english.peopledaily.com.cn
Xinhua www.xinhuanet.com/english
Far Eastern Economic Review. www.feer.com
The China Quarterly. Cambridge Journals. http://journals.cambridge.org
The Economist. www.economist.com
China labour statistical yearbook. Beijing: China Statistics Press. latest year.
China population and employment statistical yearbook. Beijing: China Statistics Press. latest year.
China statistical yearbook. Beijing: China Statistics Press. latest year.
Statistical communique of the People's Republic of China on the 2007 national economic and social development. China Statistics Press, 2008.
Women and men in China. Facts and figures 2007. Department of Social, Science and Technology. National Bureau of Statistics, 2008
Wen Jiabao. Report to the 11th National People's Congress http://npc.people.com.cn/
China Internet Network Information Center. *Statistical survey report on internet development in China.* Annual publication. www.cnnic.cn
CSM Media Research. *Rating China.* www.csm.com.cn
International Institute of Strategic Studies (IISS). *The military balance. 2004–2005.* Oxford: Oxford University Press. 2004.
United Nations Development Programme (UNDP). *China human development report.* United Nations.2005.
US Census Bureau. International database www.census.gov
World Bank. *The world development indicators.* Washington: World Bank. annual publication.

Secondary sources

Alden C. *China in Africa.* London: Zed Boooks, 2008.
Benewick. R et al, editors. *Asian politics in development.* London & Portland. Oregon: Frank Cass, 2003.
Blecher M. *China against the tide.* New York: Continuum, 2003.
Brady A-M. *Marketing dictatorship: propaganda and thought work in contemporary China.* Lanham: Rowman and Littlefield Publishers. Maryland. 2007.
Brodsgaard KE and Zheng Y, editors. *The Chinese Communist Party in reform.* London and New York: Routledge, 2006.
Buzan B and Foot R. *Does China matter?* London & New York: RoutledgeCurzon, 2004.
China Environment Series. Woodrow Wilson Center. Washington DC.
Deng Y. *China's struggle for status.* Cambridge: Cambridge University Press, 2008.
Donald SH and Gammack JG. *Tourism and the branded city: film and identity on the Pacific Rim.* Aldershot: Ashgate. 2007.
Fewsmith J. *China since Tiananman: from Deng Xiaoping to Hu Jintao.* Cambridge: Cambridge University Press, 2008.
Foot R. *Rights beyond borders.* Oxford: Oxford University Press, 2000.
Gladney DC. *Dislocating China: Muslims, minorities and other subaltern subjects.* Chicago: University of Chicago Press, 2004.
Goodman DSG, editor. *The new rich in China: future rulers. present lives.* London: Routledge, 2008.
Goodman DSG. Qinghai and the emergence of the West: nationalities, communal interaction, and national integration. *The China Quarterly.* 178. 2004, June.
Hearn. AH. *China and Latin America: the social foundations of a global alliance.* Durham. N.C: Duke University Press, 2009.
Holbig H and Ash R. *China's accession to the World Trade Organization: national and international perspectives.* London: RoutledgeCurzon, 2002.
Howell J editor. *Governance in China.* Lanham, Maryland: Roman and Littlefield, 2004.
Huang Y. *Capitalism with Chinese characteristics: entrepreneurship and the state.* Cambridge: Cambridge University Press, 2008.
Jeffreys E, editor. *Sex and sexuality in China.* London; New York: RoutledgeCurzon, 2006.
Jeffreys E and Sigley G, editor. *China and governmentality.* Special issue of *Economy and Society* (35 (4). London: Routledge, 2006.
Krug, B, editor. *China's rational entrepreneurs: the development of the new private business sector.* London: RoutledgeCurzon, 2004.
Kynge J. *China shakes the world.* London: Phoenix, 2007.
Lai CP. *Media in Hong Kong: press freedom and political change. 1967-2005.* London: Routledge, 2007.
Lipman JN. *Familiar strangers: a history of Muslims in northwest China.* Washington: University of Washington Press, 1997.
Mackay J and Eriksen M. *The tobacco atlas.* Geneva: World Health Organization, 2002.
Mackerras C et al, editors. *Dictionary of the politics of the People's Republic of China.* London and New York: Routledge. 1998.
Marton AM. *China's spatial economic development: regional transformation in the Lower Yangzi Delta.* London: Routledge, 2000.
Maclaren AE, editor. *Chinese women: living and working.* London: RoutledgeCurzon, 2004.
Nolan. P. *China at the crossroads.* Cambridge: Polity Press, 2004.
O'Brien J and Palmer M. *The atlas of religion.* London: Earthscan, 2007.
Pan PP. *The struggle for the soul of new China.* New York: Simon and Schuster, 2008.
Saich T. *Governance and politics of China.* Basingstoke and New York: Palgrave Macmillan, 2004.
Shambaugh. D. *Modernizing China's military.* Berkeley and London: University of California Press, 2004.
Shambaugh. D, editor. *The modern Chinese state.* Cambridge: Cambridge University Press, 2000.
Shambaugh. D. *China's Communist Party: atrophy and adaptation.* Berkeley, California: University of California Press, 2008.
Shih C-Y. *Negotiating ethnicity in China: citizenship as a response to the state.* Routledge, 2002.
Smith D. *The Atlas of war and peace.* London: Earthscan; New York: Penguin, 2003.
White G et al. *In search of civil society.* London: Macmillan, 1997.
Zweig D. *Internationalizing China: domestic interests and global linkages.* Cornell University Press, 2002.
Zhu Y. *Television in post-reform China: serial dramas. Confucian leadership. and the global television market.* London: Routledge, 2008.

INDEX